Lecture Notes in Mathematics

Edited by A. Dold and B. Eckmann

1072

Franz Rothe

Global Solutions of Reaction-Diffusion Systems

Springer-Verlag
Berlin Heidelberg New York Tokyo 1984

Author

Franz Rothe
Lehrstuhl für Biomathematik, Universität Tübingen
Auf der Morgenstelle 28, 7400 Tübingen, Federal Republic of Germany

AMS Subject Classifications (1980): 35B35, 35B40, 35B45, 35B65, 35K55, 92A17, 92A40

ISBN 3-540-13365-8 Springer-Verlag Berlin Heidelberg New York Tokyo
ISBN 0-387-13365-8 Springer-Verlag New York Heidelberg Berlin Tokyo

Library of Congress Cataloging in Publication Data Rothe, Franz, 1947 – Global solutions of reaction-diffusion systems. (Lecture notes in mathematics; 1072) Bibliography: p. Includes index. 1. Differential equations, Partial-Numerical solutions. 2. Differential equations, Parabolic-Numerical solutions. 3. Biomathematics. I. Title. II. Series: Lecture notes in mathematics (Springer-Verlag); 1072. QA3.L28 no. 1072 [QA377] 515.3'53 84-13887
ISBN 0-387-13365-8 (U.S.)

Printing and binding: Beltz Offsetdruck, Hemsbach / Bergstr.
2146/3140-543210

Preface

This monograph is motivated by some problems from Mathematical Biology. Although there exists an extensive literature about nonlinear parabolic differential equations, none of the known results could be used to prove global existence of solutions for the reaction-diffusion systems considered in this monograph. In this situation, I gave an ad hoc proof of global existence for the one-dimensional reaction-diffusion system with reaction $A + B \rightleftharpoons C$ subject to the mass action law. Afterwards it turned out that the method used could be generalized and applied to other problems as well. For the time being, the subject is not yet exhausted. Further interesting examples from applications are needed in order to build a substantial theory which is not just an unnecessarily abstract disguise of some specific problem. The author hopes that this monograph will be useful to stimulate research in this direction.

I take the opportunity to thank Prof. Dr. K.-P. Hadeler from the Lehrstuhl of Biomathematik, University of Tübingen, for his intensive personal and scientific support as well as my colleagues Dr. W. Ebel and Dr. H. Munz for reading the manuscript and giving much constructive critisism. With this help, I hope, the manuscript became more rigorous and more readable at the same time.

May 1984 Franz Rothe

Contents

Introduction

Reaction-diffusion equations have found a considerable amount of interest in recent years. They arise naturally in a variety of models from theoretical physics, chemistry and biology. Some survey and further references can be found in the books and articles by Diekmann and Temme [10], Fife [13], Fitzgibbon and Walker [17], Henry [24] and Mottoni [43].
To give some common examples for reaction-diffusion systems in applications, we mention dynamics of nuclear reactors (Kastenberg and Chambré [26], Rumble and Kastenberg [58], Mottoni and Tesei [42], example in part II), chemical reactions in distributed media and combustion theory (dissertation Ebel [11], example in part II), ecological interactions in spatially distributed populations (Alikakos [1], Conway and Smoller [9], Mimura and Murray [40], Rothe [57], example in part II), morphogenetic models (Maginu [35], Meinhardt and Gierer [39], Rothe [53], example in part II), motion of bacteria by chemotaxis (Keller and Odell [28], Rascle [50]), nerve pulse propagation (Lopes [34], Schwan [60], example in part II), models from neurophysiology (an der Heiden [23]) and population genetics (Fisher [16], Peletier [47], Rothe [54]).
It should be stressed that this choice of examples influenced by the author's preferences is of course incomplete and subjective.

Reaction-diffusion systems can give rise to a number of interesting phenomena like e.g. threshold behavior, multiple steady states and hysteresis, spatial patterns, moving fronts or pulses and oscillations. The study of these phenomena needs a variety of different methods from many areas of mathematics as for example numerical analysis, bifurcation and stability theory, semigroup theory, singular perturbations, phase space or topological methods and many others. To get a more complete survey the reader is refered to the monographs of Diekmann and Temme [10] or Fife [13].

The present work is concerned with some very common aspects of reaction-diffusion equations, namely global existence or blow-up in finite time, respectively for the solutions of the initial-boundary value problem, furtheron construction of a priori estimates and investigation of the asymptotic behavior for large time. These aspects are studied in a general framework in part I and for some examples in part II. Of course, specific applications motivate in many cases a more complete study of special examples. We mention e.g. the work of Ebel [11] about transport through biological membranes.

Existence and uniqueness of classical solutions of reaction-diffusion equations for smooth initial data local in time in standard (see e.g. Friedman [19] or Ladyzenskaja [32] for results in Hölder theory and Henry [24] or Kielhöfer [29] for a more general setting in Banach spaces). As for ordinary differential equations it may happen that the solution blows up at a finite time (see e.g. Ball [6] or Kielhöfer [29]). This time will be referred to as "maximal existence time". A priori estimates for some Hölder norm or some suitable norm defined using fractional powers of the Laplacian (Henry [24], Kielhöfer [29]) exclude such a behavior. For the examples occuring in applications it is difficult to get such strong a priori estimates. For reaction-diffusion equations of the type $u_t - \Delta u = F(x,t,u)$ where the nonlinearity F does not depend on the gradient ∇u it turns out that a priori estimates of the uniform norm are sufficient to exclude a blow-up. Even to get such estimates is a nontrivial problem.

A natural approach is to study at first the pure reaction system and ignore diffusion totally. If the solutions of this system are bounded, one might guess that the solutions of the original reaction-diffusion system are bounded as well. As pointed out already by Turing [64] and later by Prigogine [48] this guess is wrong even for linear systems. In fact Turing's argument shows that diffusion with different diffusion coefficients for different components may destabilize equilibria which are stable for the pure reaction system. By the way, this gives rise to a mechanism of pattern formation studied by Maginu [35], Mottoni [43], Rothe [53] in a very simple model, by Gierer and Meinhardt [38,39] in a biologically more realistic model as well as by Mimura and Murray [40] for some ecological models.

Nevertheless it seems quite clear for these and many other examples that local destabilization does not give rise to globally unbounded solutions. A well-known simple but very rough method to get global a priori bounds is the use of invariant rectangles or upper and lower solutions (see e.g. Chueh et.al. [8], Conway and Smoller [9], Kuiper [31] or Pao [45, 46]). This method fails for most of the quoted examples unless strong saturation terms are introduced into the equations. The generalization, the method of invariant regions (see e.g. Amann [3] or Weinberger [66]), is restricted to the case of equal diffusion coefficients for all components. This is often just the uninteresting case. Thus many differential equations from applications admit neither invariant rectangles nor any invariant convex sets. Then all methods mentioned above fail.

A typical example is the simple chemical reaction $A + B \rightleftharpoons C$ (see example in part II). This example suggested by Ebel [11] has been investigated in great detail in her dissertation. It has been the starting point of this work.

Thus more involved methods to get a priori bounds must be developped. The general procedure consists of two steps. In a first step some functional of the solution is shown to be bounded. For this initial step, similar to the construction of Lyapunov functionals, there does not exist a general method, but many examples can be handled by individual tricks. In a second step it is shown that a priori bounds for the functional constructed above imply uniform a priori bounds. For this second step a general satisfactory formulation can be given. This method was used by Alikakos [1,2] in some special cases. Rothe [56] contains some simplifications and generalizations, which are continued in the present work.

Part I contains a general existence theory for reaction-diffusion equations including the improvement of a priori bounds.
Part II gives various examples. The primary a priori estimates are constructed by quite different methods. Then the results and methods of part I are used to construct uniform a priori bounds. Hence one gets global classical solutions. In some cases a Lyapunov functional is used furtheron to get sharp results about the asymptotic behavior.

Although we cannot give a complete review of the literature concerned with global existence of solutions for semilinear parabolic equations of the type $u_t - \Delta u = F(x,t,u,\nabla u)$, we mention the classical work of Friedman [20] and Sobolevski [59] as well as the recent work of Amann [3], Kielhöfer [29,30], Pao [45,46] and von Wahl [65].
Sobolevski [59] proves local existence of solutions of nonlinear parabolic differential equations in an abstract Banach space and already uses the fractional powers A^{α} of the generator A of a semigroup. Friedman [20] gives an application of these results of Sobolevski to concrete nonlinear parabolic initial-boundary value problems. His paper is indeed closely related to this monograph. It already contains the relation (1) of p.6 between the exponent γ for the growth of the nonlinearity and the exponent q from the primary L_q-estimate of the solution. On the other hand, Friedman restricts himself to smooth initial data u_o with $A^{\alpha}u_o \in E$ for some $\alpha > 0$ and exponents $q \in [1,\infty]$. He does not give an **explicit** formula for the final uniform estimate as it is derived in this monograph by the "feedback argument".

(see e.g. p.69, Lemma 19 formula (214) below).

The work of Pao [45,46] relies totally on monotonicity arguments and the maximum principle.

Amann [3] assumes the existence of invariant regions as they were introduced in a more elemantary fashion by Weinberger [66] and formulates the evolution equation and the invariance assumption in Banach spaces.

Kiehöfer's work [30] is inspired by the Navier-Stokes equations. There a first a priori estimate comes from energy conservation. Explicitly his basic assumption (FO) means $\int_\Omega Fu\,dx \leq K \int_\Omega (u^2+(\nabla u)^2)\,dx$. Also this paper uses a very abstract setting of functional analysis. In principle the methods ressemble the "bootstrapping argument" explained below. The "feedback" is not used, but instead a remarkable Lemma comes in.

This "Lemma of Kielhöfer" has in some sense the same spirit as the methods of this monograph and seems to be even sharper.

It uses both the smoothing properties of the semigroup and some primary a priori estimate e.g. from energy conservation.

Von Wahl's work [65] is inspired by minimal surfaces. Roughly spoken von Wahl shows the following: if the difference of two solutions of a reaction-diffusion system can a priori be uniformly estimated by the uniform norm of the difference of the corresponding initial- and boundary values, then a global classical solution exists. In case of minimal surfaces, this assumption can be fulfilled.

As a common point, the more recent paper just mentioned include dependence of the nonlinearity F on the gradient ∇u, e.g. Amann allows the nonlinearity $(\nabla u)^{2-\varepsilon}$, Kielhöfer $u(\nabla u)$ from the Navier-Stokes equations and von Wahl even $(\nabla u)^2$. In this sense they are more general than the present monograph, in which dependence of the nonlinearity F on the gradient ∇u will not be considered. On the other hand, this monograph handles very weak primary a priori estimates occuring in applications. This seems indeed to be the main progress.

Part I <u>Existence and A Priori Estimates for Reaction-Diffusion Equations</u>

The setting of part I is adapted to applications like the examples in part II. In some cases (e.g. models of nerve pulse propagation, Schwan [60] or ecological models, Rothe [57]) there arise systems of reaction--diffusion equations coupled with ordinary differential equations. For such systems the solutions do not smooth out the initial data. To include this type of systems and to avoid regularity considerations at the very beginning, we manage by working with mild solutions. They are defined as solutions of a corresponding integral equation. This setting is used throughout in part I.

The Theorems of part I can be applied in different ways. Most simply the whole reaction-diffusion system can be considered globally without special regard to the individual components. Then some obvious modifications including vector-valued functions u,c and F in the Theorems below are needed. We will not take this point of view, because it turns out to be too rough.Indeed, in many cases one obtains sharper results by applying the Theorems below to the diffusion equations for the individual components. Focussing attention to the equation for one component we have to care of the other components, because they may appear in the nonlinearity F. There we treat them as "weight function" $c = c(x,t)$ and assume only very little information about them as e.g. some L_q-bound given by the primary a priori estimate of the solution of the whole system.

Our main goal is to derive a uniform a priori estimate from some primary estimate of a L_r-norm. To this end we will persue two methods.

The first method may be characterized as "smoothing for short time". The basic arguments are contained in Lemma 22 and 23. The idea is to con-construct local solutions with initial data in the space L_r and values in L_∞ for positive time. Repeating this procedure at different initial times yields an L_∞-estimate. Theorem 1, 2, 5(iii) and 6(ii) are based on this argument.

The second method may be characterized as "bootstrap and feedback". The basic arguments are given in Lemma 18 and 19. In this reasoning existence of solutions for smooth initial data is taken to be granted from the beginning. Then the smoothing properties of diffusion as stated in Lemma 3 are used to improve the primary a priori estimate by means of the mild formulation. This kind of argument is well known as "bootstrapping" (see e.g. Massatt [36]). It seems to be much less known that the scope of the bootstrapping can be enlarged by a trick which we will call "feedback".

To give an example, let $r \in (0,\infty)$, $\gamma \in [1,\infty)$, $\Omega \subset R^N$, u be a smooth function $(x,t) \in \bar{\Omega} \times [0,\infty) \to u(x,t) \in [0,\infty)$ with the following properties:

$$0 \leq u \quad \text{and} \quad u_t - \Delta u \leq u^\gamma \quad \text{on } \Omega \times [0,\infty); \quad u(x,0) = 0 \quad \text{for all } x \in \Omega \quad \text{and}$$

$$\|u\|_r := \sup\{\|u(.,t)\|_r \mid t \in [0,\infty)\} < \infty.$$

It is straightforward to show that the condition

$$(\gamma - 1)N/(2r) < 1 \tag{1}$$

is necessary to guarantee that the bootstrap argument can sharpen the L_r-estimate. It is quite surprising that (1) is also sufficient. One does not need the additional assumption $\gamma \leq r$ of Massatt [36]. To see this take the difficult case that (1) holds but $\gamma > r$. Then one cannot "start" the bootstrap argument, because there is no L_1-estimate available for the nonlinear term u^γ. This case can nevertheless be handled by the use of an interpolation inequality and the feedback argument. The simple example just explained is contained e.g. in Rothe [56]. The main idea is contained in the following diagram (where $u \in E_r$ means that $\|u\|_r := \sup\{\|u(.,t)\|_r \mid t \in [0,\infty)\} < \infty$):

$$u \in E_r \to \boxed{\begin{array}{c} \|u^\gamma\|_p \leq \\ \|u\|_r^{r/p}\|u\|_\infty^{\gamma - r/p} \end{array}} \to u^\gamma \in E_p \to \boxed{\begin{array}{c} u_t - \Delta u \leq u^\gamma, \ 0 \leq u, \\ p \in [1,\infty], \ p > N/2 \\ \text{implies} \\ \|u\|_\infty \leq K \|u^\gamma\|_p \end{array}} \to u \in E_\infty$$

$$\boxed{\text{feedback with } \gamma - r/p < 1}$$

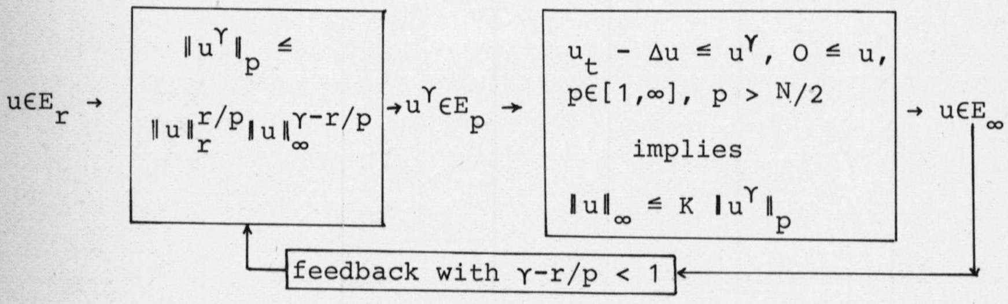

We have the chain of inequalities $\|u\|_\infty \leq K\|u^\gamma\|_p \leq K\|u\|_r^{r/p}\|u\|_\infty^{\gamma - r/p}$

and hence by feedback $\|u\|_\infty < \infty$ implies $\|u\|_\infty \leq (K\|u\|_r^{r/p})^{\frac{1}{1-(\gamma - r/p)}}$.

This "feedback" brings several advantages. Firstly, as we have just seen, the condition $\gamma \leq r$ can be dropped, Only (1) is the important assumption. Even the case $r < 1$ can be included. Thus the primary a priori estimate needs not to be a norm, but can be a L_r-estimate for some $r \in (0,\infty)$ and for sublinear equations even a weaker estimate as considered in Theorem 7. Finally, at least for bounded initial data, it is easy to give an explicit formula for the bound of $\|u\|_\infty$.
In the present paper we generalize the feedback argument to initial data $u_0 \in L_{p_0}$ and the case of a priori estimates involving space- and time-integrals. The general initial data can only be handled by introducing weighted norms allowing singularities near $t = 0$ and repeating the argument in a finite chain in different spaces. The details are given in Lemma 16,18,19 and 20 and the results are formulated in Theorem 5(i)(ii) and 6(i).

A third method to derive a priori estimates given by Alikakos [1,2] should be mentioned. In this method the diffusion equation is tested by powers of the solution u. Then in an iterative procedure estimates of the norms $\|u\|_{2^\nu r}$ are given for $\nu = 0,1,2,...$ successively. Each step uses the Gagliardo-Nirenberg inequality. This method is very complicated, because one has to care for the limit $\nu \to \infty$. For the equations considered in this monograph it does not yield better results (see e.g. Rothe [56] for details). It turns out to be of advantage in the porous media equation, where no formulation as an integral equation is available (see Alikakos [2]). The feedback argument used in this paper cuts down the whole process of Alikakos like summing up a geometrical series. As an additional advantage, the Gagliardo-Nirenberg inequality is replaced by the elementary interpolation between L_p-spaces.

It may be be interesting to note that none of the two methods used in this paper to derive a priori estimates turns out to be better. The differences in the results obtained can be seen in Theorem 5 and 6. Theorem 5(iii)(iv) and 6(ii) are based on the first method, whereas Theorem 5(i)(ii),6(i) and 7 use the second method.

Some remarks should be made concerning the existence proofs. We want to note that they were not the starting point of this work. Indeed, our main goal is the construction of uniform a priori bounds. As done in Rothe [56] one could restrict oneself to smooth initial data. Then existence of local solutions for the initial-boundary value problem is assured by the classical results of Friedman [19] or Ladyzenskaja [32].

In this monograph we look for more complete results. Theorem 1,2 and 3 prove existence of mild solutions. Less smooth initial data $u_0 \in L_{p_0}$ are included. There arise difficulties concerning the choice of a suitable function space for the Picard-Lindelöf iteration. As noted by Weissler [67], this difficulty can be handled by introducing weighted L_p-norms which allow a singularity of the solution u at t = 0. To prove that the solutions constructed are in L_∞ for all t > 0 a chain of different function spaces must be used. The details are given in Lemma 12,13 and 14. The knowledge of some a priori bounds can be used to prove the existence of solutions even for initial data which are not admitted in Theorem 1 or 2. This reasoning is used in Theorem 3,4 and 6.

Many authors as e.g. Amann [3], Henry [24], Kielhöfer [29,30], Weissler [67] have formulated existence theorems in a more abstract way using two Banach spaces E and $E_\alpha \subset E$. The basic reason for this setting is that the semigroup S(t) maps E into the smaller space E_α but the nonlinearity F maps E_α into the larger space E. But all authors mentioned above except Weissler [67] restrict the initial data to the smaller space E_α . Weissler includes initial data $u_0 \in E$. The essential assumptions about the semigroup S(t) and the nonlinearity F are

$$\|S(t)u\|_{E_\alpha} \le Kt^{-a}\|u\|_E \quad \text{and} \quad \|F(u)-F(v)\|_E \le l(\|u\|_{E_\alpha}+\|v\|_{E_\alpha})\|u-v\|_{E_\alpha}$$

and some relation between the number $a \in R$ and the growth of the function $l = l(u)$ for $u \to \infty$ as e.g. $\int_1^\infty u^{-1/a} l(u)\, du < \infty$.

In the present paper no attempt is made to show that the mild solutions constructed are differentiable solutions of an abstract differential equation in some Banach space. This is no serious drawback, since for positive time the mild solutions are indeed classical solutions. A detailed proof is given in part II, review of standard theorems.

The arguments of Theorem 1,2,3 and 4 use a global Lipschitz condition which restricts the growth of the derivatives $\partial F/\partial u$ of the nonlinearity F for $u \to \infty$ very strongly. This is too restrictive for applications from part II. For the derivation of a priori estimates in Lemma 19,20 and 23 it is sufficient to restrict only the growth of the nonlinearity F itself. For nonnegative solutions u, it is even enough to have an upper bound for F. With regard to the applications we avoid the global Lipschitz condition in Theorem 5,6 and 7. Nevertheless existence of solutions is shown. To this end, a local Lipschitz condition is assumed, which ensures existence of solutions for bounded initial data.

Since uniform bounds imply bounds in Hölder spaces (Lemma 21), existence can be shown by a Peano argument using compactness. Of course one cannot show uniqueness by this method.

It remains to ask in what sense the results given here are optimal. Weissler [67] shows that the diffusion equation

$$u_t - \Delta u = u|u|^{\gamma-1} \quad \text{on } \Omega\times[0,\infty), \quad \text{where } \Omega \subset R^N \text{ is a bounded domain,}$$

does not have any local solution for certain initial data $u_o \in L_{p_o}$, $u_o \geq 0$ if $(\gamma-1)N/(2p_o) > 1$. Hence the regularity assumption

$$N/(2q_1)+(1/q_2)+(\gamma-1)N/(2p_o) \leq 1 \qquad (54) \text{ or } (L)$$

of Theorem 1 and 2 cannot be relaxed if one has to prove existence of a local mild solution for initial data $u_o \in L_{p_o}$.

Inclusion of the equality sign in (54) needs special considerations. In that case, which we call the "limit case", the existence proof given in Theorem 1 must be modified. The smoothing properties of the linear semi-group S(t) for the individual initial data $u_o \in L_{p_o}$ as considered in Lemma 4 must be exploited. The contraction argument used in Theorem 2 takes a space the norm of which is adapted to the individual initial data u_o.

In the limit case with equality sign in (54) the construction of uniform a priori estimates becomes difficult. For the first method (smoothing for small time) Theorem **5(iii)** gives a weak result. In Lemma 24, one does not get a functional relation between the estimates of $\|u\|_{p_o}$ and $\|u\|_\infty$ as in Lemma 23. Indeed Lemma 23 and Theorem 6 break down in the limit case. The second method to construct uniform a priori bounds cannot be used in the limit case. It may be interesting that the Lemma of Kielhöfer (appendix of [30]) still yields some weaker results. This method is applied in part II to the Brusselator and yields uniform estimates for space dimension N = 4, whereas the results of part I can only be used for space dimension N = 1,2 and 3.

If (54) is violated, there is no hope to get uniform a priori estimates. This is illustrated by the following example given by Haraux and Weissler [22]:

Let $\gamma \in (1,\infty)$ such that $1 < N(\gamma-1)/2 < \gamma+1$ and $1 = (\gamma-1)N/(2p_o)$

Then the Cauchy problem $u(x,0) \equiv 0$, $u_t - \Delta u = u|u|^{\gamma-1}$ for $x \in R^N$, $t \in (0,\infty)$ has a nontrivial, quite smooth, positive solution u such that

$$\|u(.,t)\|_r = t^{-N(1/p_0 - 1/r)/2} \|u(.,1)\|_r > 0 \quad \text{for all } t \in (0,\infty),\ r \in [1,\infty].$$

Obviously, we have the a priori bound

$$\max\{\|u(.,t)\|_r \mid t \in [0,T]\} < \infty \quad \text{for all } T > 0 \text{ if } (\gamma-1)N/(2r) \geqq 1,$$

but no uniform a priori bound.

Finally we mention a peculiarity. The trajectory may be bounded in some space L_r for small r violating (1). Then the following strange situation may arise: The solution u explodes after a finite time $T_{max} < \infty$ in the sense of Theorem 5 (ii), but nevertheless the trajectory is bounded in the sense that

$$\sup\{\|A^\alpha u(.,t)\|_p \mid t \in [0,T_{max})\} < \infty$$

if $\alpha \in (0,1)$, $p \in [1,r)$ satisfy $\alpha + N(\gamma/r - 1/p)/2 < 1$. (For the definition of the fractional powers A^α of the generator $A = -\Delta$ see Henry [24]) In this situation the trajectory

$$\{u(.,t) \in L_p \mid t \in [T_{max}/2, T_{max})\} \text{ is precompact in the space } L_p.$$

Hence the following limit set ω is nonvoid and compact in L_p:

$$\omega = \{v \in L_p \mid \text{there exists a sequence } t_n \to T_{max} \text{ such that } \lim_{n\to\infty} \|u(.,t_n) - v\|_p = 0\}$$

It is impossible to exploit this fact for the extension of the solution u to some larger interval $[0,T_{max}+\varepsilon)$. To see this, note that by assumption $(\gamma-1)N/(2p) > 1$. Hence by a result of Weissler [67] mentioned above there may not exist local solutions for initial data $v \in \omega \subset L_p$ taken from the limit set ω.

The scalar equation $u_t - \Delta u = u|u|^{\gamma-1}$ has been studied by Ball [6] in a bounded domain and by Weissler [68] in the whole space R^N. Figuereido et.al. [14] give conditions for the existence of bounded equilibrium solutions.

Basic Notations and Definitions

Let the positive integer N denote the space dimension. For simpler notation we define $n = N/2$. Thus the half-integer n is half of the space dimension. Denote by $x = (x_1, x_2, \ldots, x_N)$ a generic point in R^N. Let Ω be a bounded domain in R^N whose boundary $\partial\Omega$ is an (N-1)-dimensional $C^{2+\alpha}$-manifold for some $\alpha \in (0,1)$ such that Ω lies locally on one side of $\partial\Omega$.

For $p \in [1,\infty]$ let $L_p(\Omega)$ be the well-known Banach space of measurable functions $u: x \in \Omega \to u(x) \in R$ endowed with the norm

$$\|u\|_p = \left[\frac{1}{|\Omega|}\int_\Omega |u(x)|^p dx\right]^{1/p} \quad \text{for } p \in [1,\infty)$$

$$\|u\|_\infty = \sup\text{ess}\{|u(x)|; x \in \Omega\}$$

Since Ω is supposed to be bounded we get for all $p,q \in [1,\infty]$ with $p \leq q$

$$\|1\|_p = 1 \quad \text{and} \quad L_q \subset L_p \quad \text{and} \quad \|u\|_p \leq \|u\|_q \quad \text{for all } u \in L_q.$$

Denote by $(x,t) = (x_1, x_2, \ldots, x_N, t)$ a generic point in $R^N \times R$. For parabolic equations functions of space x and time t will be considered.

Let $T \in (0,\infty]$, $\delta \in [0,\infty)$, $p, p_1, p_2 \in [1,\infty]$. Define the interval $I(T)$ by $I(T) = (0,T]$ for $T < \infty$ or $I(T) = (0,\infty)$ for $T = \infty$, respectively.
Let the function $m: t \in [0,\infty) \to m(t) \in [0,1]$ be given by

$$m(t) = \min(1,t).$$

Let $E_{p,\delta,T}$ be the Banach space of measurable functions

$u: (x,t) \in \Omega \times I(T) \to u(x,t) \in R$ such that (i)(ii) hold:

(i) $u(.,t) \in L_p$ for all $t \in I(T)$ without exceptional null set

(ii) The norm $\|u\|_{p,\delta,T} := \sup\{m(t)^\delta \|u(.,t)\|_p \mid t \in I(T)\}$ is finite.

Let $L_{p_1,p_2,T}$ be the Banach space of measurable functions

$u: (x,t) \in \Omega \times I(T) \to u(x,t) \in R$ endowed with the (finite) norm

$$\|u\|_{p_1,p_2,T} = \sup\left\{\left[\int_{t_1}^{t_2}\left(\frac{1}{|\Omega|}\int_\Omega |u(x,t)|^{p_1} dx\right)^{p_2/p_1} dt\right]^{1/p_2} \,\middle|\, \begin{matrix} t_{1,2} \in [0,T], \\ 0 \leq t_2 - t_1 \leq 1 \end{matrix}\right\}$$

The cases $p_1, p_2, T = \infty$ are included by obvious modifications.

Let $L_{p_1,p_2} = L_{p_1,p_2,\infty}$ and $\|u\|_{p_1,p_2} = \|u\|_{p_1,p_2,\infty}$. The choice of the norms $\|u\|_{p,\delta,\infty}$ and $\|u\|_{p_1,p_2}$ is adapted to the study of the asymptotic behavior for $t \to \infty$. The functional $\|u\|_{p_1,p_2,T}$ with $p_1,p_2 \in (0,\infty]$ occurs, too.

For $\alpha \in (0,1)$ let $C(\bar{\Omega})$ and $C^\alpha(\bar{\Omega})$ be the Banach spaces of continuous and α-Hölder continuous functions $u: x \in \bar{\Omega} \to u(x) \in R$ endowed with the norms

$$\|u\|_\infty = \max\{|u(x)|;\ x \in \bar{\Omega}\} \quad \text{and}$$

$$\|u\|_{C^\alpha} = |u|_\infty + \max\{|u(x)-u(y)|/|x-y|^\alpha \mid x,y \in \bar{\Omega}\}.$$

For $p \in N$, $\alpha \in (0,1)$ let $C^p(\bar{\Omega})$ and $C^{p+\alpha}(\bar{\Omega})$ be the Banach spaces of continuously differentiable and α-Hölder continuously differentiable functions $u \in C(\bar{\Omega})$ endowed with the usual norms.

For $\mu, \alpha \in (0,1)$ let $C^\mu([0,T],C^\alpha(\bar{\Omega}))$ be the Banach space of α-Hölder continuous functions $u: t \in [0,T] \to u(.,t) \in C^\alpha(\bar{\Omega})$ endowed with the norm

$$\|u\| = \max\{\|u(.,t_1)\|_{C^\alpha} + \|u(.,t_1)-u(.,t_2)\|_{C^\alpha}/|t_1-t_2|^\mu \mid t_{1,2} \in [0,T]\}.$$

This part treats semilinear parabolic initial-boundary value problems of the following type:

$$u_t + Lu = F(x,t,u) \qquad \text{for all } x \in \Omega,\ t > 0 \tag{1}$$

$$bu + \delta \frac{\partial u}{\partial \rho} = 0 \qquad \text{for all } x \in \partial\Omega, t > 0 \tag{1a}$$

$$u(x,0) = u_o(x) \qquad \text{for all } x \in \Omega \tag{1b}$$

Here (1a) is called the boundary condition and (1b) the initial condition. In the following the initial-boundary value problem (1)(1a)(1b) will simply be denoted by IBP.
Now we give more detailed assumptions about the nonlinear function F (modelling e.g. chemical reactions) and the elliptic operator L as well as the boundary conditions (modelling e.g. diffusion).

For the function $F: (x,t,u) \in \Omega \times [0,\infty) \in R \to F(x,t,u) \in R$ the following assumptions occur:

(FO) For all $u \in R$ the function $F(.,.,u): (x,t) \in \Omega \times [0,\infty) \to F(x,t,u) \in R$ is measurable

Only assumption (F0) will be made throughout. We will always state explicitly, which of the following assumptions (F1) through (F7) are needed. Assumptions (F1)(F2)(F3) still depend on the quantities $\gamma \in [1,\infty)$ and $q_1, q_2 \in [1,\infty]$, whereas assumptions (F4)(F5)(F6) depend on $q_1, q_2 \in [1,\infty]$. These quantities will be specified in every case later. Assumptions (F2)(F5)(F7) will be used if one considers nonnegative solutions $u \geq 0$. Now we state assumptions (F1) through (F7):

There exists a function $c \in L_{q_1,q_2}$ such that for almost all $(x,t) \in \Omega \times [0,\infty)$ the following holds, respectively:

(F1) $$|F(x,t,u)| \leq c(x,t)(1+|u|)^{\gamma} \quad \text{for all } u \in R$$

(F2) $$F(x,t,u) \leq c(x,t)(1+u)^{\gamma} \quad \text{for all } u \in [0,\infty)$$

(F3) $$|F(x,t,u)-F(x,t,v)| \leq c(x,t)(1+|u|+|v|)^{\gamma-1}|u-v| \quad \text{for all } u,v \in R$$

There exists a function $c \in L_{q_1,q_2}$ and an increasing function $\Gamma: u \in [0,\infty) \to \Gamma(u) \in [0,\infty)$ such that for almost all $(x,t) \in \Omega \times [0,\infty)$:

(F4) $$|F(x,t,u)| \leq c(x,t)\Gamma(|u|) \quad \text{for all } u \in R$$

(F5) $$F(x,t,u) \leq c(x,t)\Gamma(u) \quad \text{for all } u \in [0,\infty)$$

(F6) $$|F(x,t,u)-F(x,t,v)| \leq c(x,t)\Gamma(|u|+|v|)|u-v| \quad \text{for all } u,v \in R$$

(F7) $$F(x,t,0) \geq 0$$

Next we need some definitions and properties concerning the diffusion term. We consider the linear initial-boundary value problem

$$u_t + Lu = 0 \quad \text{for all } x \in \Omega,\ t \in (0,\infty) \tag{2}$$

$$b(x)u(x,t) + \delta\frac{\partial u}{\partial \rho}(x,t) = 0 \quad \text{for all } x \in \partial\Omega,\ t \in (0,\infty) \tag{1a}$$

$$u(x,0) = u_0(x) \quad \text{for all } x \in \Omega \tag{1b}$$

Here the elliptic operator L is formally defined by

$$Lu := -\sum_{j,k=1}^{N} \frac{\partial}{\partial x_k} a_{jk}(x) \frac{\partial}{\partial x_j} u + \sum_{k=1}^{N} b_k(x) \frac{\partial}{\partial x_k} u + c(x)\, u \tag{3}$$

The following assumptions (L0) through (L5) are made throughout. Let $\alpha \in (0,1)$ be some number.

(L0) $a_{jk}, b_k \in C^{1+\alpha}(\bar{\Omega})$; $b, \rho_k \in C^{1+\alpha}(\partial\Omega)$; $c \in C^{\alpha}(\bar{\Omega})$

Let the operator L be uniformly elliptic i.e. $a_{kj} = a_{jk}$ and there exists a constant $M > 0$ such that

(L1) $$\frac{1}{M}\sum_{j=1}^{N} y_j^2 \leq \sum_{j,k=1}^{N} a_{jk}(x)y_jy_k \leq M\sum_{j=1}^{N} y_j^2 \quad \text{for all } (y_1 \ldots y_N) \in R^N, x \in \bar{\Omega}.$$

There exists a constant $c_o \in R$ (on which assumption (L5) below indeed imposes further restrictions) such that

(L2) $c(x) \geq c_o$ for all $x \in \bar{\Omega}$.

Let (n_k) denote the outer normal unit vector at the boundary $\partial\Omega$. We consider two types of boundary conditions:

(L3) Let one of the following cases (δ0) or (δ1) occur:

(δ0) $\delta = 0$ and $b(x) \equiv 1$ for all $x \in \partial\Omega$

(δ1) $\delta = 1$ and $b(x) \geq 0$, $\rho_j(x) = \sum_{k=1}^{N} a_{jk}(x)n_k(x)$ for all $x \in \partial\Omega$

Hence $\partial/\partial\rho = \sum_{j=1}^{N} \rho_j(x)\partial/\partial x_j$ is the derivative in the conormal direction, which is nontangential and pointing outward by assumption (L1). The boundary condition (1a) is in the cases (δ0)(δ1), respectively:

(δ0) $u = 0$ on $\partial\Omega$; Dirichlet boundary condition

(δ1) $bu + \partial u/\partial\rho = 0$ on $\partial\Omega$; Neumann- or third type boundary cond.

Trivially the assumptions (L0) through (L4) are fulfilled if $L = -\Delta$ with Dirichlet- or Neumann boundary conditions. Only these cases are relevant for the applications in part II. Assumption (L5) can only be explained later on p.23.

The linear initial-boundary value problem (1a)(1b)(2) can be treated as an ordinary differential equation in several Banach spaces as e.g. $L_p(\Omega)$, for $p \in [1,\infty]$, $C(\bar{\Omega})$ or $C^{\alpha}(\bar{\Omega})$ with $\alpha \in (0,1)$. First we define the operator A_o:

$$D(A_o) = \{u \in C^2(\bar{\Omega}) \mid u \text{ satisfies the boundary condition (1a)}\}$$

$$A_o u = L u \quad \text{for all } u \in D(A_o).$$

The numbers $p, q, \nu, \mu, q_1, q_2, \alpha, \gamma$ etc. will be call exponents.

By the letter K we will denote constants depending only on the domain Ω, the operator A_o and the exponents. Other quantities which K may depend on (as e.g. time t) are always mentioned explicitly in the form K(t).

Lemma 1

Let $p\in(1,\infty)$. The operator A_o is closable in the Banach space $L_p(\Omega)$. The closure will be denoted by A_p. The operators A_p are generators of analytic semigroups $S_p(t)$ in the spaces $L_p(\Omega)$.

Let $p,q\in(1,\infty)$ and $p \leq q$. Then the following holds:

$$A_q \subset A_p \quad \text{i.e.} \quad D(A_q) \subset D(A_p) \quad \text{and} \quad A_p u = A_q u \quad \text{for all } u \in D(A_q) \tag{4}$$

$$S_q \subset S_p \quad \text{i.e.} \quad S_p(t)u = S_q(t)u \quad \text{for all } t \in [0,\infty),\ u \in L_q \subset L_p \tag{5}$$

Let $u_o \in L_p$ for some $p\in(1,\infty)$. By convenient redefinition on a set $\Sigma(t) \subset \overline{\Omega}$ of measure zero for each $t \in [0,\infty)$, the function

$$u(x,t) = (S_p(t)u_o)(x)$$

can be chosen continuous for $(x,t) \in \overline{\Omega}\times(0,\infty)$. Then even the derivatives u_t, $u_{x_i t}$, $u_{x_i x_k}$ exist and are α-Hölder continuous for $(x,t) \in \overline{\Omega}\times(0,\infty)$.
The function u satisfies the differential equation (2) and the boundary condition (1a) in the classical sense for $(x,t) \in \overline{\Omega}\times(0,\infty)$.
The initial data u_o are assumed in the sense

$$\lim_{t\to 0} \|u(.,t) - u_o\|_p = 0 \tag{6}$$

Proof

By Friedman [18]p.74, Theorem 19.1 the operator A_o satisfies

$$\|u\|_{W_p^2} \leq K_1(\|A_o u\|_p + \|u\|_p) \quad \text{for all } u \in D(A_o) \tag{7}$$

where the constant K_1 is independent of u. Here W_p^2 is the Sobolev space of functions with generalized derivatives of second order in $L_p(\Omega)$.
Hence the operator A_o is closable in L_p and the closure A_p satisfies

$$D(A_p) \subset W_p^2(\Omega) \quad \text{and} \quad \|u\|_{W_p^2} \leq K_1(\|A_p u\|_p + \|u\|_p) \quad \text{for all } u \in D(A_p) \tag{8}$$

Let the operator B be defined as the restriction of the operator A_p to the space $L_q \subset L_p$, i.e.

$$Bu = A_p u \quad \text{for all } u \in D(B) = D(A_p) \cap L_q.$$

Then $A_o \subset B \subset A_p$. Since B is closed in L_q, we get $A_q \subset B \subset A_p$ and hence (4).
By Friedman [18] p.101, the resolvent set $\mathbb{C} \setminus \sigma(A_p)$ contains an angular sector $\Sigma = \{z \in \mathbb{C} \mid |\arg(z-k)| > \pi/2 - \varphi\}$ for some $k \in R$, $\varphi \in (0,\pi/2)$. The operator A_p generates an analytic semigroup $S_p(t)$ which is given explicitly by the formula

$$S_p(t) := e^{-A_p t} = \frac{1}{2\pi i}\int_\gamma e^{-zt}(A_p - z)^{-1}dz. \tag{9}$$

Here γ is an infinite curve along the boundary of the sector Σ e.g. $\gamma: s \in (-\infty,\infty) \to z(s) \in \mathbb{C}\setminus\sigma(A_p)$ given by

$$z(s) = k-1 + e^{i(\pi/2-\varphi)}s \quad \text{for } s \in [0,\infty)$$

$$z(s) = k-1 - e^{-i(\pi/2-\varphi)}s \quad \text{for } s \in (-\infty,0].$$

Let $z \in \sigma(A_p) \cap \sigma(A_q)$. By definition of A_p, A_q we get

$$(A_p - z)^{-1}v = (A_q - z)^{-1}v \quad \text{for all } v \in R(A_o - z), \text{ the range of } A_o - z.$$

Since $R(A_o - z)$ is dense in L_p and L_q we get by closure

$$(A_p - z)^{-1}v = (A_q - z)^{-1}v \quad \text{for all } v \in L_q \subset L_p.$$

Now the representation (9) of the semigroups yields (5).

Let $u_o \in L_p$ and $u(x,t) = (S_p(t)u_o)(x)$. By Friedman [18] p.105 we have

$$u(.,t) \in D(A_p^m) \quad \text{and} \quad t^m \|A_p^m u(.,t)\|_p \le K_m e^{-kt}\|u_o\|_p$$
$$\text{for all } m = 0,1,2,\ldots \quad \text{and } t \in (0,\infty) \tag{10}$$

where the constants K_m depend only on m and A_p.
Especially (8) and (10) for m = 0,1 imply

$$\|u(.,t)\|_{W_p^2} \le K_2(1+t^{-1})e^{-kt}\|u_o\|_p \quad \text{for all } t \in (0,\infty). \tag{11}$$

Applying (8) once again to $v = A_p u(.,t) \in D(A_p)$ and using (10) for m = 1,2 yields

$$\|A_p u(.,t)\|_{W_p^2} \le K_3(t^{-1} + t^{-2})e^{-kt}\|u_o\|_p \quad \text{for all } t \in (0,\infty). \tag{12}$$

Let $\alpha \in (0,1)$ be the number occuring in assumption (LO) and choose $\nu \in (1,2)$, $p \in (1,\infty)$ such that $1+\alpha < \nu < 2-N/p$. Firstly we prove the Lemma for (large) p restricted in this way. Let $t \in (0,\infty)$ be arbitrary. The Sobolev inequality (see e.g. Henry [24])

$$C^{\nu}(\bar{\Omega}) \subset W_p^2(\Omega) \quad \text{and} \quad |u|_{C^{\nu}} \leq K_4 \|u\|_{W_p^2} \quad \text{for all } u \in W_p^2(\Omega) \tag{13}$$

together with (8) and (10) for m = 1,2 implies

$$\left.\begin{aligned} u(.,t) &\in D(A_p) \subset W_p^2(\Omega) \subset C^{\nu}(\bar{\Omega}) \\ A_p u(.,t) &\in D(A_p) \subset W_p^2(\Omega) \subset C^{\nu}(\bar{\Omega}) \end{aligned}\right\} \quad \text{for all } t \in (0,\infty)$$

Since $u(.,t) \in D(A_p)$, there exists a sequence u_n in $D(A_0)$ such that

$\lim_{n\to\infty} \|u(.,t)-u_n\|_p + \|A_p u(.,t)-A_0 u_n\|_p = 0$. Hence by (8) and (13) also

$$\lim_{n\to\infty} \|u(.,t)-u_n\|_{W_p^2} + \|u(.,t)-u_n\|_{C^{\nu}} = 0.$$

Since $u_n \in D(A_0)$ satisfy the boundary condition (1a) in the classical sense, this shows that $u(.,t)$ satisfies the boundary condition (1a) in the classical sense for all $t \in (0,\infty)$.

Fix $t \in (0,\infty)$; $\alpha \in (0,1)$, $\nu \in (1,2)$ and $p \in (1,\infty)$ are as above. Choose $z \in \mathbb{R} \setminus \sigma(A_p)$ and let $w = (A_p - z)u(.,t)$. Then the elliptic boundary value problem

$$(A_0 - z)v = w, \quad v \in D(A_0) \tag{14}$$

can be treated in the classical Schauder theory since $w \in C^{\nu}(\bar{\Omega}) \subset C^{\alpha}(\bar{\Omega})$ and the coefficients occuring in the operator A_0 and the boundary conditions are assumed to be smooth enough, namely in $C^{1+\alpha}$ besides c which is in C^{α}. For the classical Schauder theory of elliptical problem, see e.g. Fiorenza [15]. Thus (14) has a classical solution $v \in C^{2+\alpha}(\bar{\Omega})$, which is unique in $C^{2+\alpha}(\bar{\Omega})$ and satisfies the Schauder estimate

$$\|v\|_{C^{2+\alpha}} \leq K_5(z)\|w\|_{C^{\alpha}}. \tag{15}$$

Hence $u(.,t)-v \in L_p(\Omega)$ satisfies $(A_p - z)(u(.,t)-v) = 0$.
Since $z \in \mathbb{R} \setminus \sigma(A_p)$ this implies

$$u(.,t) = v \quad \text{in the space } L_p(\Omega)$$

and after a convenient redefinition of $u(.,t)$ on a null set $\Sigma(t) \subset \bar{\Omega}$ even

$u(x,t) = v(x)$ for all $x \in \bar{\Omega}$. (Dependence of w and v on t is suppressed).

Hence $u(.,t) \in C^{2+\alpha}$ and $A_p u(.,t) \in C^{\nu}(\bar{\Omega})$ for all $t \in (0,\infty)$.

It remains to show that the differential equation (2) is satisfied classically. Since $S_p(t)$ is an analytic semigroup in L_p, we know that for all $t\in(0,\infty)$ the function $u(.,t) = S_p(t)u_o$ is infinitely differentiable with respect to the variable t in the L_p-sense and $u_t = A_p u$. Hence Taylor's formula yields for all $t\in(0,\infty)$, $h\in(-t,\infty)$:

$$u(.,t+h) = u(.,t) + A_p u(.,t)h + \int_0^1 (1-s)A_p^2 u(.,t+sh)ds\, h^2.$$

Using (8) and (13) we arrive at the estimate

$$\|u(.,t+h)-u(.,t)-A_p u(.,t)h\|_{C^\nu} \leq K_6 h^2 \max_{0\leq s\leq 1} \|(A_p^2+A_p^3)u(.,t+sh)\|_p.$$

Hence (10) implies

$$\left\|\frac{u(.,t+h)-u(.,t)}{h} - A_p u(.,t)\right\|_{C^\nu} \leq K_7 h(1+t^{-3})e^{-kt}\|u_o\|_p.$$

Taking the limit $h \to 0$, this shows that the time derivative $u_t(.,t)$ exists in the space $C^\nu(\bar{\Omega})$ for all $t\in(0,\infty)$. Summarizing the results we get

$$u_t(.,t) = A_p u(.,t) \in C^\nu(\bar{\Omega});\ u_{tx_i}(.,t) \in C^{\nu-1}(\bar{\Omega}) \subset C^\alpha(\bar{\Omega}) \text{ and}$$
$u_{x_i x_k}(.,t) \in C^\nu(\bar{\Omega})$ for all $t \in (0,\infty)$, which proves the assertions.

We remove the restriction imposed on p.17 on the exponent p. Let $p\in(1,\infty)$ be arbitrary and $u_o \in L_p(\Omega)$. For some $l \in N$ we choose finite sequences (t_i) and (p_i) with $t_i\in(0,\infty)$ and $p_i\in(1,\infty)$ for $i = o,1,2...l$ such that

$$p_o = p,\quad 1/p_i < 1/p_{i-1} \leq 1/p_i + 2/N \quad \text{for } i = 1...l \text{ and}$$

$$1 + \alpha < \nu < 2 - N/p_l.$$

The Sobolev inequalities (13) and for $i = 1...l$

$$L_{p_i}(\Omega) \subset W^2_{p_{i-1}}(\Omega) \text{ and } \|u\|_{p_i} \leq K_8\|u\|_{W^2_{p_{i-1}}} \quad \text{for all } u \in W_{p_{i-1}} \tag{16}$$

together with (11) imply inductively:

$$u(.,t_o) = S_{p_o}(t_o)u_o \in W^2_p \subset L_{p_1}$$

$$u(.,t_o+t_1) = S_{p_1}(t_1)u(.,t_o) \in W_{p_1} \subset L_{p_2}$$

$$u(.,t_0+t_1+t_2) = S_{p_2}(t_2)u(.,t_0+t_1) \in W^2_{p_2} \subset L_{p_3}$$

$$u(.,t_0+t_1+\ldots+t_{l-1}) = S_{p_{l-1}}(t_{l-1})u(.,t_0+t_1+\ldots+t_{l-2}) \in W^2_{p_{l-1}} \subset L_{p_l}$$

$$u(.,t_0+t_1+\ldots+t_l) = S_{p_l}(t_l)u(.,t_0+t_1+\ldots t_{l-1}) \in W^2_{p_l} \subset C^\nu(\bar{\Omega})$$

Hence $u(.,t) \in C^\nu(\bar{\Omega})$ for all $t\in(0,\infty)$. Thus the arguments above prove the assertions of the Lemma.
Finally (6) follows since $S_p(t)$ is an analytic and hence also a strongly continuous semigroup. Thus the Lemma is proved.

It turns out to be useful to include the less regular cases $p = 1$ and ∞.

Lemma 2
(i) The semigroups $S_p(t)$ are preserving positivity:

$u_0 \in L_p(\Omega)$, $u_0(x) \geq 0$ for almost all $x \in \Omega$ implies

$(S_p(t)u_0)(x) \geq 0$ for all $x \in \bar{\Omega}$, $t \in (0,\infty)$

(ii) $\sup \{ \|S_p(t)u\|_\infty \mid t\in[0,T]\} \leq K(T)\|u\|_\infty$ for all $u \in L_\infty$, $T\in[0,\infty)$. (17)

If the constant in assumption (L2) satisfies $c_0 = 0$, then (17) holds with $K(T) = 1$.
Define the operators $S_\infty(t)$ as the restriction of $S_p(t)$ to the space $L_\infty(\Omega)$. These operators are well defined (independent of p) and continuous from $L_\infty(\Omega)$ to $L_\infty(\Omega)$. Furtheron, since

$$S_\infty(t_1+t_2) = S_\infty(t_1)S_\infty(t_2) \quad \text{for all } t_1,t_2 \in [0,\infty),$$

they define a formal semigroup in $L_\infty(\Omega)$. This is not a strongly continuous semigroup.

(iii) $\sup \{ \|S_p(t)u\|_1 \mid t\in[0,T]\} \leq K(T)\|u\|_1$ for all $p\in(1,\infty)$, $u \in L_p$, $T\in[0,\infty)$

The operators $S_p(t)$ can be uniquely extended to continuous operators $S_1(t)$ from $L_1(\Omega)$ to $L_1(\Omega)$. The operators $S_1(t)$ form a strongly continuous semigroup in the space $L_1(\Omega)$.

(iv) Let $p\in[1,\infty]$ including $p = 1, \infty$. Let $u_0 \in L_p(\Omega)$. Then

$$\lim_{t\to 0} \|S_p(t)u_0 - u_0\|_p = 0 \qquad (6)$$

holds if and only if the initial data u_o satisfy the following condition

(A) For $p < \infty$ no restriction;
For $p = \infty$, $u_o \in C(\bar{\Omega})$ and in case (δ0) of Dirichlet boundary conditions additionally
$u_o \in C_o(\bar{\Omega}) := \{u_o \in C(\bar{\Omega}) \mid u_o(x) = 0 \text{ for all } x \in \partial\Omega\}$

Proof

(i) We call initial data u_o regular, if (a)(b)(c) hold.

(a) $u_o \in C^{2+\alpha}(\bar{\Omega})$; (b) $b(x)u_o(x) + \partial u_o/\partial\rho(x) = 0$ for all $x \in \partial\Omega$

(c) In case (δ0) of Dirichlet boundary conditions $L u_o \in C_o(\bar{\Omega})$

By Ladyzenskaja [32] p.320, Theorem 5.2, 5.3 for regular initial data there exists a unique classical solution $u = u(x,t)$ of the linear initial-boundary value problem (2)(1a)(1b). Furtheron u_t and $u_{x_i x_k}$ are α-Hölder continuous on the closed set $(x,t) \in \bar{\Omega}\times[0,\infty)$.
By uniqueness for semigroups this classical solution must be identical with the solution constructed by formula (9). Hence $u(x,t)=(S_p(t)u_o)(x)$.
By the weak maximum principle (Protter and Weinberger [49]) the assertion of (i) holds in the case of classical solutions. Since the set of regular initial data is dense in $L_p(\Omega)$ for any $p\in[1,\infty)$, the assertion (i) follows.

(ii) Let $e = e(x,t)$ be the classical solution of the differential equation (2) with the following initial- and boundary conditions in case (δ0),(δ1) respectively:

In case (δ0): $e(x,0) = 1$ for all $x \in \bar{\Omega}$
$e(x,t) = 1$ for all $x \in \partial\Omega$, $t \in [0,\infty)$
In case (δ1): $e(x,0) = 1$ for all $x \in \bar{\Omega}$
$b(x)e(x,t) + \frac{\partial e}{\partial\rho}(x,t) = 0$ for all $x \in \partial\Omega$, $t \in [0,\infty)$

Let $u = u(x,t)$ be the solution of the initial-boundary value problem (2)(1a)(1b) for regular initial data u_o. Then the comparison theorem for parabolic equations (see e.g. Protter and Weinberger [49] or review of standard theorems part II) implies

$$|u(x,t)| \leq e(x,t)\|u_o\|_\infty \quad \text{for all } x \in \bar{\Omega},\ t \in [0,\infty) \tag{18}$$

Let some initial data $u_o \in L_\infty$ be given. We choose $\nu \in (0,2)$ and $p\in(N,\infty)$ such that $\nu < 2-N/p$ and a sequence u_{om} of regular initial data u_{om} such that

$$\|u_{om}\|_\infty \leq \|u_o\|_\infty \quad \text{and} \quad \lim_{m\to\infty} \|u_{om}-u_o\|_p = 0. \tag{19}$$

Then (18) and (11)(13), respectively imply for all $t \in (0,\infty), m \in N$:

$$\lim_{m\to\infty} \|S_p(t)(u_{om}-u_o)\|_{C^\nu} = 0 \quad \text{and} \quad \|S_p(t)u_{om}\|_\infty \leq \|e(.,t)\|_\infty \|u_{om}\|_\infty \tag{20}$$

Now (19)(20) prove the assertion of (ii) with $K(T)=\sup\{\|e(.,t)\|_\infty | t\in[0,T]\}$. If the constant in assumption (L2) satisfies $c_o \geqq 0$, the comparison theorem shows $e(x,t) \leqq 1$ for all $(x,t) \in \bar{\Omega}\times[0,\infty)$ and hence $K(T) \leqq 1$.

(iii) We argue by duality. The formal adjoint L^* of the operator L defined by formula (3) is

$$L^* v = - \sum_{j,k=1}^{N} \frac{\partial}{\partial x_j} a_{jk}(x) \frac{\partial}{\partial x_k} v - \sum_{k=1}^{N} \frac{\partial}{\partial x_k} b_k(x)\, v + c(x)\, v. \tag{21}$$

In cases $(\delta 0)$,$(\delta 1)$ respectively we need the boundary conditions

$$\begin{aligned} &\text{In case } (\delta 0): && v = 0 \quad \text{on } \partial\Omega \\ &\text{In case } (\delta 1): && (b + \sum_{k=1}^{N} n_k b_k) v + \sum_{j,k=1}^{N} n_j a_{jk} \frac{\partial}{\partial x_k} v = 0 \quad \text{on } \partial\Omega \end{aligned} \tag{1a}^*$$

We define the operator B_o by

$D(B_o) = \{v \in C^2(\bar{\Omega}) \mid v$ satisfies the boundary condition $(1a)^*\}$

$B_o v = L^* v$ for all $v \in D(B_o)$.

Let $(.,.)$ denote the usual scalar product in $L_2(\Omega)$. Some elementary partial integrations show that

$$(A_o u,v) = (u,B_o v) \quad \text{for all } u \in D(A_o),\ v \in D(B_o). \tag{22}$$

By Lemma 1 the operator B_o is closable in the space $L_2(\Omega)$. Let B_2 be the closure of B_o in $L_2(\Omega)$. Then (22) implies by closure

$$(A_2 u,v) = (u,B_2 v) \quad \text{for all } u \in D(A_2),\ v \in D(B_2). \tag{23}$$

By Friedman [18]p.101 the operators A_2 and B_2 are both generators of analytic semigroups. For some $k \in R$, $\varphi \in (0,\pi/2)$ there exists an angular sector $\Sigma = \{z \in \mathbb{C} \mid |\arg(z-k)| > \pi/2 - \varphi\} \subset \mathbb{C} \setminus [\sigma(A_2) \cup \sigma(B_2)]$. For all $z \in \Sigma$ we know that $R(A_2-z) = R(B_2-z) = L_2(\Omega)$ (see Friedman [18] p.77 Th.19.3). Hence (23) implies

$$(f,(B_2-z)^{-1}g) = ((A_2-z)^{-1}f,g) \quad \text{for all } f,g \in L_2,\ z \in \Sigma. \tag{24}$$

The representation (9) of analytic semigroups implies

$$(f,e^{-B_2 t}g) = (e^{-A_2 t}f,g) \quad \text{for all } f,g \in L_2,\ t \in [0,\infty) \tag{25}$$

Applying estimate (17) to the semigroup $e^{-B_2 t}$ yields

$$\|(e^{-A_2 t}f,g)\| \le \|f\|_1 \|e^{-B_2 t}g\|_\infty \le K(T)\|f\|_1\|g\|_\infty \quad \text{for all } f,g \in L_2, t\in[0,T].$$

Since $L_1^*(\Omega) = L_\infty(\Omega)$ (see e.g. Segal and Kunze [61] p.156) this implies

$$\|e^{-A_2 t}f\|_1 \le K(T)\|f\|_1 \quad \text{for all } f \in L_2,\ t \in [0,T],$$

which proves (iii) since L_2 is dense in the space $L_1(\Omega)$.

(iv) For $p\in(1,\infty)$ the assertion (6) is clear since $S_p(t)$ is a strongly continuous semigroup in the space $L_p(\Omega)$.
Let $p = 1$ or ∞, $u_o \in L_p$ and assume that assumption (A) holds. Then there there exists a sequence u_{om} of regular initial data approximating u_o in the sense of (19). Since $u_m(x,t) = (S_p(t)u_{om})(x)$ are classical solutions of the initial-boundary value problem (2)(1a)(1b) on the closed domain $(x,t) \in \bar{\Omega}\times[0,\infty)$ (see Ladyzenskaja[32] p.320, Th. 5.2, 5.3) we get

$$\lim_{t\to 0} \|S_p(t)u_{om} - u_{om}\|_\infty = 0 \quad \text{for all } m \in N. \tag{26}$$

Hence we can estimate by the triangle inequality and (ii)(iii):

$$\|S_p(t)u_o-u_o\|_p \le \|S_p(t)u_{om}-u_{om}\|_p + \|S_p(t)(u_{om}-u_o)\|_p + \|u_{om}-u_o\|_p$$
$$\le \|S_p(t)u_{om}-u_{om}\|_p + K_9\|u_{om}-u_o\|_p \quad \text{for all } t\in[0,1].$$

Let $\varepsilon > 0$ be given. By (19) there exists some $m \in N$ such that the second term can be estimated by ε. Hence (26) implies

$$\limsup_{t\to 0} \|S_p(t)u_o - u_o\|_p \le \varepsilon.$$

Since $\varepsilon > 0$ is arbitrary this implies the assertion (6).

Conversely let $p = \infty$ and assume that (6) holds. By Lemma 1 the function $S_\infty(t)u_o \in C(\bar{\Omega})$ satisfies assumption (A) for all $t\in(0,\infty)$. Hence (6) implies that $u_o \in C(\bar{\Omega})$ satisfies assumption (A), too.
Thus the Lemma is proved.

Remark
After this work was nearly finished, the author heard of the paper [4] by Amann. It contains an excellent presentation of Lemma 1,2 and 3 even

in the case of more general oblique boundary conditions, proves the optimal rate of decay $e^{-\lambda t}$ in (30) of Lemma 3 and shows that $S_1(t)$ is indeed an analytic semigroup in $L_1(\Omega)$.
Clearly by (iv), $S_\infty(t)$ is not a strongly continuous semigroup in $L_\infty(\Omega)$. Nevertheless, this semigroup is useful in the existence proofs e.g. in Lemma 9 or Theorem 1 p.111. Recent results of Steward [62] show that the restriction of $S_\infty(t)$ to the spaces $C_0(\bar{\Omega})$ or $C(\bar{\Omega})$ for boundary conditions (δ0) or (δ1), respectively, indeed is an analytic semigroup.

Since the embedding $W_p^2(\Omega) \subset L_p(\Omega)$ is compact for all $p \in (1,\infty)$, estimate (8) implies that the resolvent $(A_p - z)^{-1}$ is compact for all $z \in \sigma(A_p)$. Hence the spectrum $\sigma(A_p)$ is a pure point spectrum. Let $z \in \sigma(A_p)$, $v \in L_p$ be an eigenvalue and an eigenfunction, hence $A_p v = zv$. We choose the numbers α, ν and the finite sequence (p_i) $i = 0...l$ as on p.18. Then estimate (8) and the Sobolev inequality (16) imply for $i = 1...l$:

$$\|v\|_{p_i} \leq K_8 \|v\|_{W^2_{p_{i-1}}} \leq K_{10}(\|A_{p_{i-1}} v\|_{p_{i-1}} + \|v\|_{p_{i-1}}) \leq K_{10}(1+|z|)\|v\|_{p_{i-1}} \tag{27}$$

By the same argument estimate (8) and Sobolev inequality (13) imply

$$\|v\|_{C^\nu} \leq K_{11}(1+|z|)\|v\|_{p_l}. \tag{28}$$

Let $z_1 \in \mathbb{C} \setminus \sigma(A_p)$. Then the equation $(A_p - z_1)v = (z - z_1)v$ can be looked at as an elliptic boundary value problem as (14). Hence the Schauder estimate (15) implies

$$\|v\|_{C^{2+\alpha}} \leq K_5(z)\|v\|_{C^\alpha}. \tag{29}$$

Now (27)(28)(29) show that all eigenfunctions of the operator A_p are regular, indeed $v \in C^{2+\alpha}(\bar{\Omega})$. Hence the spectrum of the operators A_p is independent of $p \in (1,\infty)$. By Friedman [18] p. 77, Th.19.3 the resolvent set $\mathbb{C} \setminus \sigma(A_p)$ contains a left half plane. Actually, Amann [4] gives even more detailed information about the spectrum. $\sigma(A) = \sigma(A_p)$ consists of the principal eigenvalue $\lambda = \min \operatorname{Re} \sigma(A)$ and a disjoint part contained in the set $\{z \mid \operatorname{Re}(z-\lambda) \geq \varepsilon,\ |\arg(z-\lambda)| \leq \pi/2 - \varepsilon\}$ for some $\varepsilon > 0$. The principal eigenvalue λ is nondegenerate and the principal eigenfunction can be chosen nonnegative.

For the investigation of the asymptotic behavior of the semigroup for $t \to \infty$ we shall assume

(L5) $\quad \min \operatorname{Re} \sigma(A) \equiv \lambda > 0.$

To this end it is sufficient to choose the constant c_0 in assumption (L2) such that $c_0 \geq 0$ and $c_0 > 0$ in case of Neumann boundary conditions.

For the nonlinear problem like (1) considered later assumption (L5) imposes no restriction. It can be fulfilled by convenient choice of the nonlinear function $F(x,t,u)$ and the function $c(x)$ in (3).

Assumptions (L1) through (L5) will be made throughout in part I.

As shown e.g. by Henry [24] p.24, the fractional powers A_p^α are well defined for all $p \in (1,\infty)$ and $\alpha \in (0,1)$. They will be used in the regularity considerations of Lemma 21 and review of standard theorems, part II. It is possible to define also semigroups $S_\nu(t)$ in the Hölder spaces $C^\nu(\bar{\Omega})$. In general they will not be strongly continuous semigroups. To see this, note that $\lim_{t\to 0} \|S_\nu(t)u_o - u_o\|_{C^\nu} = 0$ can only hold if u_o is in the $C^\nu(\bar{\Omega})$-closure of the set of regular functions as defined on p.20. Nevertheless a theory using formula (9) from the construction of analytic semigroups has been developped by Kielhöfer [29]. An additional singularity occurs for $t = 0$ in estimates as e.g.

$$\|A_\nu^\alpha S_\nu(t)u\|_{C^\nu} \le K_{12} t^{-(\alpha+\nu+\varepsilon)} \|u\|_{C^\nu}.$$

This construction brings some advantages in the case of unbounded domains, but will be avoided in this work. In Lemma 21 we will get estimates of the Hölder norm via the continuous embeddings $W_p^2 \subset C^\nu(\bar{\Omega})$ for $\nu < 2-N/p$ and $D(A_p^\alpha) \subset C^\nu$ for $\nu < 2\alpha - N/p$. Furtheron the classical Schauder estimates as e.g. (15) or Ladyzenskaja [32] p.320, Theorem 5.2,5.3 are used.

The next Lemma is fundamental for the whole work. It contains estimates of the semigroup $S_p(t)$ considered as an operator between different spaces. Since by Lemma 1 we have the inclusions $L_p \subset L_1$ and $S_p(t) \subset S_1(t)$ for all $p \in [1,\infty]$, we can drop the index p and simply write $S(t)$ for the semigroup. For simpler notation of the expressions involving the exponents N,p,q_1,q_2,r_1,r_2 etc, we define half of the space dimension N: $n = N/2$ and drop some brackets! The denominator after a fraction bar / consists only of a single number if no brackets occur. Hence we write e.g.

$$N/(2p) = n/p \; ; \quad (N/2)(1/p-1/q) = n/p-n/q \; ;$$

$$N/(2q_1)+(1/q_2)+(\gamma-1)N/(2p) = n/q_1+1/q_2+(\gamma-1)n/p.$$

Recall that $m(t) = \min\{1,t\}$ for all $t \in [0,\infty)$.

If $u = u(x,t)$ is a function $(x,t) \in \Omega \times (0,T) \to u(x,t) \in R$, we denote by $u(t)$ or $u(.,t)$ the function $x \in \Omega \to u(x,t) \in R$. Whether "$.,$" is dropped or not is more or less chance.

Lemma 3

(i) Let $p,q\in[1,\infty]$ and $p \le q$. Then

$$\|S(t)u\|_q \le K_{13} m(t)^{-(n/p-n/q)} e^{-\lambda t} \|u\|_p \quad \text{for all } u \in L_p,\ t \in (0,\infty) \tag{30}$$

(ii) Let $\nu\in(0,2)$ and $\rho\in(0,\infty)$. Then

$$\|S(t)u\|_{C^\nu} \le K_{14} m(t)^{-(\nu/2+\rho)} e^{-\lambda t} \|u\|_\infty \quad \text{for all } u \in L_\infty,\ t \in (0,\infty) \tag{31}$$

Remarks

The exponential decay $e^{-\lambda t}$ is actually given by the principal eigenvalue λ in (L5) p.23. This optimal result relying on the maximum principle was obtained only recently by Amann [4].

The semigroup $S(t)$ defines a bounded operator from L_p to L_q. We write

$$\|S(t)u\|_q \le \|S(t)\|_{q,p} \|u\|_p \quad \text{for all } t\in(0,\infty),\ u \in L_p.$$

Reed and Simon [52] call these semigroups "hypercontractive".

Proof

(i) First we assume $\alpha = n/p-n/q \in [0,1]$. By the Gagliardo-Nirenberg inequality (Friedman [18] p.27, Th.10.1) we get for $p\in(1,\infty)\setminus\{N/2,N\}$

$$\|S(t)u\|_q \le K_{15} \|S(t)u\|^{\alpha}_{W^2_p} \|S(t)\|^{1-\alpha}_p. \tag{32}$$

Hence the continuous embedding $D(A_p) \subset W^2_p$ stated in (8) implies

$$\|S(t)u\|_q \le K_{16} (\|A_p S(t)u\|_p + \|S(t)u\|_p)^{\alpha} \|S(t)u\|^{1-\alpha}_p. \tag{33}$$

Since A_p is the generator of an analytic semigroup, assumption (L5) and Friedman [18] p.101, Theorem 2.1 imply that for some $\bar{\lambda}$ we have

$$\|S(t)u\|_p \le K_{17} e^{-\bar{\lambda} t} \|u\|_p \quad \text{for all } u \in L_p,\ t \in [0,\infty). \tag{34}$$

By Amann [4] we can choose $\bar{\lambda} = \lambda$ given by (L5). Now (10) with $m = 1$, (34) and **the** semigroup property $S(t_1+t_2) = S(t_1)S(t_2)$ imply

$$\|A_p S(t)u\|_p \le K_{18} m(t)^{-1} e^{-\lambda t} \|u\|_p \quad \text{for all } u \in L_p,\ t \in (0,\infty). \tag{35}$$

Now inserting (34)(35) in (33) yields the assertion (30) under the additional assumptions $n/p-n/q\in[0,1]$ and $p\in(1,\infty)\setminus\{N/2,N\}$.

As in the proof of Lemma 2(iii) we argue by duality in order to remove the restrictions for $p\in[1,\infty]$. Let $p,p',q,q'\in[1,\infty]$, $p \le q$, $p \notin\{1,N/2,N\}$ and $1/p+1/p' = 1/q+1/q' = 1$. By formula (25) application of (30) [with the restrictions for p and q] to the adjoint semigroup yields

$$\begin{aligned} |(S(t)u,v)| &= |(u,e^{-B_2 t}v)| \le \|u\|_{q'}\|e^{-B_2 t}v\|_q \quad \text{for all } t \in (0,\infty), \\ &\le K_{19}m(t)^{-(n/p-n/q)}e^{-\lambda t}\|u\|_{q'}\|v\|_p \quad u\in L_2\cap L_{q'};\ v\in L_2\cap L_p. \end{aligned} \tag{36}$$

Since $p'< \infty$ we know that $L^*_{p'}(\Omega) = L_p(\Omega)$ (see e.g. Segal and Kunze [61] p.154,156). Hence (36) implies

$$\|S(t)u\|_{p'} \le K_{19}m(t)^{-(n/q'-n/p')}e^{-\lambda t}\|u\|_{q'} \quad \text{for all } u \in L_{q'}, t \in (0,\infty)$$
$$\text{if } n/q'-n/p'\in[0,1] \text{ and } p'\in(1,\infty)\setminus\{N/(N-1),N/(N-2)\}.$$

Note that no restriction for $q'\in[1,\infty]$ occurs in this estimate. Hence the semigroup property $S(t_1+t_2) = S(t_1)S(t_2)$ is used to remove all restrictions for $p'\in[1,\infty]$. Thus estimate (30) is proved.

For illustration we give a more simple and direct proof in the special case $L = -\Delta$ with Dirichlet boundary condition (1a)(δ0). Let $v\colon (x,t)\in R^N\times[0,\infty) \to v(x,t)\in R$ be the solution of the Cauchy problem

$$v_t - \Delta v = 0 \qquad \text{on } R^N\times(0,\infty);$$
$$v(x,0) = \begin{cases} |u_o(x)| & \text{for } x \in \Omega \\ 0 & \text{for } x \in R^N\setminus\Omega. \end{cases}$$

The solution v can be calculated explicitly:

$$v(x,t) = (4\pi t)^{-N/2}\int_\Omega \exp\left[-\frac{(x-y)^2}{4t}\right]|u_o(y)|dy.$$

Hence Young's inequality for convolutions (see e.g. Reed and Simon [52] p.29) shows

$$\|v(.,t)\|_q \le \|\varphi\|_r\|v(.,0)\|_p \le K_{20}\, t^{-(n/p-n/q)}\|v(.,0)\|_p \quad \text{where} \tag{37}$$

$$\varphi(y) = (4\pi t)^{-n}\exp[-y^2/(4t)] \quad \text{and} \quad 1/r = 1+1/q-1/p.$$

The comparison theorem (see e.g. review of standard theorems, part II) implies

$$|u(x,t)| \le v(x,t) \quad \text{for all } (x,t) \in \bar\Omega\times[0,\infty). \tag{38}$$

Now (37)(38) imply (30) with $\lambda = 0$. Since the exponential factor $e^{-\lambda t}$ can occur in (30) only in the case of bounded domains Ω, the argument above valid for the heat equation on the whole space R^N cannot yield this exponential factor.

(ii) Choose $\alpha \in (0,1)$ and $p \in (1,\infty)$ such that $\nu/2+n/p < \alpha < \nu/2+\rho$. We use the fractional powers A_p^α of the generator A_p defined by Henry [24] p.24. By [24] p.26, Theorem 1.4.3 and (34) we have

$$\|A_p^\alpha S(t)u\|_p \leq K_{21} m(t)^{-\alpha} e^{-\lambda t} \|u\|_p \quad \text{for all } u \in L_p,\ t \in (0,\infty). \tag{39}$$

By [24] p.39, Theorem 1.6.1 we have the Sobolev inequality

$$D(A_p^\alpha) \subset C^\nu(\bar{\Omega}) \quad \text{and} \quad \|u\|_{C^\nu} \leq K_{22} \|A_p^\alpha u\|_p \quad \text{for all } u \in D(A_p^\alpha). \tag{40}$$

Combining (39) and (40) yields

$$\|S(t)u\|_{C^\nu} \leq K_{22} \|A_p^\alpha S(t)u\|_p \leq K_{23} m(t)^{-\alpha} e^{-\lambda t} \|u\|_p \quad \text{for all } u \in L_p,\ t \in (0,\infty)$$

and hence the assertion (31).
Thus the Lemma is proved.

<u>Lemma 4</u>
Let $p,q \in [1,\infty]$ and $p < q$. Let the set $\Pi \subset L_p(\Omega)$ have the following property (K):

(K) The set $k(\Pi) := \{u/|u|_p \mid u \in \Pi,\ u \neq 0\}$ is precompact in $L_p(\Omega)$

(This is fulfilled e.g. if Π is precompact and $0 \notin \overline{\Pi}$ or if Π is precompact and $\Pi = E \cap B$ where E is a linear space and $B = \{u \in L_p \mid |u|_p < 1\}$)

Then there exists a continuous nondecreasing function $g: t \in [0,\infty) \to g(t) \in [0,1]$ depending only on the exponents p,q,N, the operator A_0 and the set $\Pi \subset L_p(\Omega)$ such that (a)(b)(c) hold:

(a) $\|S(t)u\|_q \leq K_{13} g(t) m(t)^{-(n/p-n/q)} e^{-\lambda t} \|u\|_p$ for all $u \in \Pi$, $t \in (0,\infty)$. (41)

(b) $\lim_{t \to 0} g(t) = 0$

(c) The function $W = W(t)$ defined by $W(t)^{-(n/p-n/q)} = g(t) m(t)^{-(n/p-n/q)}$

satisfies $m(t) \leq W(t) \leq m(t)^{1/2}$ and $\lim_{t \to 0} W(t) = 0$.

Proof

First notice that $p < q$ implies $\sup_{0<t<1} \|S(t)\|_{q,p} = \infty$. (Otherwise $\|S(t)u\|_q \leq M\|u\|_p$ for all $u \in C^{2+\alpha}(\bar{\Omega})$, $t \in (0,1)$ and hence by taking the limit $t \to 0$ also $\|u\|_q \leq M\|u\|_p$ for all $u \in C^{2+\alpha}(\bar{\Omega})$ would follow, which is obviously impossible). Hence the semigroup property $S(t_1+t_2) = S(t_1)S(t_2)$ implies

$$\lim_{t\to 0} \|S(t)\|_{q,p} = \infty. \tag{42}$$

We define the function $h: (t,u) \in (0,\infty) \times L_p(\Omega) \setminus \{0\} \to h(t,u) \in [0,1]$ by

$$h(t,u) = \|S(t)u\|_q \|S(t)\|_{q,p}^{-1} \|u\|_p^{-1} .$$

Clearly (42) implies

$$\lim_{t\to 0} h(t,v) = 0 \quad \text{for all } v \in L_q(\Omega). \tag{43}$$

By assumption (K) the set $k(\Pi)$ is precompact in the space L_p. Since $p < \infty$ $L_q \subset L_p$ is a dense subset of L_p. Hence for any $\varepsilon > 0$ there exists a finite set $\{v_1, v_2, \dots, v_M\} \subset L_q(\Omega) \setminus \{0\}$ such that

$$\min_{1 \leq m \leq M} \left\| \frac{u}{\|u\|_p} - v_m \right\|_p \leq \varepsilon \quad \text{for all } u \in \Pi \setminus \{0\}.$$

Now we estimate for some $m \in \{1,2,\dots,M\}$ suitably choosen:

$$\begin{aligned} \left\| S(t) \frac{u}{\|u\|_p} \right\|_q &\leq \|S(t)\|_{q,p} \left\| \frac{u}{\|u\|_p} - v_m \right\|_p + \|S(t)v_m\|_q \\ &\leq \varepsilon \|S(t)\|_{q,p} + (1+\varepsilon) \|S(t)v_m\|_q \|v_m\|_p^{-1}. \end{aligned}$$

Hence we get

$$h(t,u) \leq \varepsilon + (1+\varepsilon) \max_{1 \leq m \leq M} h(t,v_m) \quad \text{for all } u \in \Pi \setminus \{0\}. \tag{44}$$

By (43) and (44) the function

$$\tilde{h}(t) = \sup \{h(s,u) \mid s \in [0,t],\ u \in \Pi,\ u \neq 0\}$$

satisfies $0 \leq \tilde{h} \leq 1$ and $\lim_{t\to 0} \tilde{h}(t) = 0$. By the definition of the functions h and $\tilde{h}$ we get

$$\|S(t)u\|_q \leq \tilde{h}(t) \|S(t)\|_{q,p} \|u\|_p \quad \text{for all } u \in \Pi,\ t \in (0,\infty).$$

Setting $g(t) = \sup\{\tilde{h}(t), m(t)^{(n/p-n/q)/2}\}$ all assertions are fulfilled by (30) and the estimate above.

Now we turn to the nonlinear initial-boundary value problem (1)(1a)(1b), which will furtheron be called IBP. In part I it will be convenient to work with mild solutions. Let $T \in (0,\infty)$ and denote by I a time interval $[0,T)$, $[0,T]$ or $[0,\infty)$.

Definition 1

By a mild solution of the IBP on the time interval I we mean a measurable measurable function $u: (x,t) \in \Omega \times I \to u(x,t) \in R$ such that (a)(b)(c)(d) hold:

(a) $u(.,t) \in L_1(\Omega)$ for all $t \in I \setminus \{0\}$

(b) $F(.,t,u(.,t)) \in L_1(\Omega)$ for all $t \in I \setminus \{0\}$ except a null set

(c) $\int_0^t \|F(.,s,u(.,s))\|_1 ds < \infty$ for all $t \in I$

(d) $u(.,t) = S(t)u_0 + \int_0^t S(t-s)F(.,s,u(.,s))ds$ for all $t \in I \setminus \{0\}$.

The integral is an absolutely converging Bochner integral in the space $L_1(\Omega)$ (see e.g. Segal and Kunze [61] p.169).

Restrictions and extensions of the time interval I will not be indicated explicitly. For simpler notation we define the functions

$$u(t): x \in \Omega \to u(t)(x) = u(x,t) \in R \quad \text{and}$$

$$\tilde{F}(t,u(t)): x \in \Omega \to \tilde{F}(t,u(t))(x) = F(x,t,u(x,t)) \in R$$

denoting by $\tilde{F}$ the Nemyzky operator

$$\tilde{F}: (t,v) \in ([0,\infty) \times L_1(\Omega)) \cap D \to \tilde{F}(t,v) = F(.,t,v(.)) \in L_1(\Omega)$$

on the natural domain of definition

$$D = \{(t,v) \in [0,\infty) \times L_1(\Omega) \mid F(.,t,v(.)) \in L_1(\Omega)\}.$$

Hence (d) can be written more simply (dropping the "$\tilde{}$"):

$$u(t) = S(t)u_0 + \int_0^t S(t-s)F(s,u(s))\,ds \quad \text{for all } t \in I. \tag{45}$$

The initial data u_0 are build in the the integral equation (45). One has to be careful in which sense the initial data u_0 are assumed. In part I we assume $u_0 \in L_{p_0}$ for some $p_0 \in [1,\infty]$ and prove that the initial data are assumed in the sense

$$\lim_{t\to 0} \| u(t) - S(t)u_o \|_{p_o} = 0. \tag{46}$$

Thus the initial data are assumed for the nonlinear initial-boundary value problem (1)(1a)(1b) "as good as" for the linear problem (2)(1a)(1b). By Lemma 2(iv) we know that

$$\lim_{t\to 0} \| S(t)u_o - u_o \|_{p_o} = 0 \tag{6}$$

holds if and only if assumption (A) p.20 is satisfied. By (6) and (46)

$$\lim_{t\to 0} \| u(t) - u_o \|_{p_o} = 0$$

holds, if the initial data u_o satisfy assumption (A) p.20.

By the semigroup property $S(t_1+t_2) = S(t_1)S(t_2)$ it is easy to show that the integral equation (45) implies

$$u(t_1+t_2) = S(t_2)u(t_1) + \int_0^{t_2} S(t_2-s)F(t_1+s,u(t_1+s))ds \tag{47}$$
$$\text{for all } t_{1,2} \text{ such that } t_1,t_2,t_1+t_2 \in I.$$

Lemma 5 and 6 give some estimates for the time-convolutions in formulas like (45).

Lemma 5

Let $p\in(1,\infty]$, $r\in[1,\infty]$, $\alpha,\beta\in[0,1)$, $\delta\in[0,\infty)$ satisfy

$$\alpha < 1 - 1/p, \quad \beta < 1 - 1/p, \quad \alpha + \beta < \delta + 1 + 1/r - 1/p. \tag{48}$$

Let a function $f \in L_p[0,T]$ be given and define the function $g: t\in[0,T] \to g(t)\in R$ by

$$g(t) := t^{\delta}\int_0^t (t-s)^{-\beta}s^{-\alpha}f(s)\, ds. \tag{49}$$

Then $g \in L_r[0,T]$ and there exists a constant $K(\alpha,\beta,\delta,p,r)$ such that

$$\| g \|_r \le K\, T^{(\delta+1+1/r-1/p-\alpha-\beta)} \| f \|_p. \tag{50}$$

Proof

Define $q\in[1,\infty)$ by $1/p+1/q = 1$. By the Hölder inequality (49) implies

$$\|g(t)\| \leq t^{\delta}\left[\int_0^t (t-s)^{-\beta q} s^{-\alpha q}\, ds\right]^{1/q} \|f\|_p$$

$$\leq t^{(\delta+1/q-\alpha-\beta)}\left[\int_0^1 (1-s)^{-\beta q} s^{-\alpha q}\, ds\right]^{1/q} \|f\|_p. \quad \text{Hence}$$

$$\|g\|_r \leq \left[\int_0^T t^{(\delta+1/q-\alpha-\beta)r}\, dt\right]^{1/r} K(\alpha,\beta,\delta,p,r)\|f\|_p$$

$$\leq K\, T^{(\delta+1/r+1/q-\alpha-\beta)}\|f\|_p.$$

Here (48) ensures convergence of the integrals. Thus the Lemma is proved.

Lemma 6

Let $p,s_1\in[1,\infty]$, $s_2\in(1,\infty]$, $\alpha,\varepsilon\in[0,1]$, $\delta\in[0,\infty)$ satisfy

$$\begin{aligned} &n/s_1 + 1/s_2 < 1 + n/p, && n/s_1 + 1/s_2 + \alpha + \varepsilon \leq 1 + n/p + \delta, \\ &1/s_2 + \alpha < 1, && 1/s_2 + \alpha + \varepsilon \leq 1 + \delta. \end{aligned} \tag{51}$$

Let $f: t\in[0,\infty) \to f(t)\in R$ be a measurable function such that

$$p_{s_2}(f) := \sup\left\{\left[\int_{t_1}^{t_2} |f(\tau)|^{s_2}\, d\tau\right]^{1/s_2} \middle|\ t_1,t_2\in[0,\infty),\ 0 \leq t_2-t_1 \leq 1\right\} < \infty.$$

Define the function $g: t\in[0,\infty) \to g(t)\in R$ by setting

$$g(t) := m(t)^{\delta}\int_0^t \|S(t-\tau)\|_{p,s_1} m(\tau)^{-\alpha} f(\tau)\, d\tau. \tag{52}$$

Then there exists a constant $K_{25}(s_1,s_2,p,\alpha,\delta,\varepsilon,N,A_0)$ such that

$$|g(t)| \leq K_{25} m(t)^{\varepsilon} p_{s_2}(f) \quad \text{for all } t \in (0,\infty). \tag{53}$$

Proof

Define $s\in[1,\infty)$ by $1/s+1/s_2 = 1$. If $p = s_1$, let $\beta = 0$. If $p > s_1$, choose $\beta\in(0,1/s)$ such that

$$n/s_1 - n/p < \beta \quad \text{and} \quad 1/s_2 + \alpha + \beta + \varepsilon \leq 1 + \delta.$$

By Lemma 3(i) there exists a constant K_{26} such that

$$|S(\tau)|_{p,s_1} \leq K_{26} m(\tau)^{-\beta} e^{-\lambda\tau}.$$

We distinguish the cases $t \leq 1$ and $t > 1$. First let $t \leq 1$. Then

$$\begin{aligned}|g(t)| &\leq K_{26}t^{\delta}\int_0^t (t-\tau)^{-\beta}\tau^{-\alpha}f(\tau)\,d\tau\\ &\leq K_{26}t^{\delta}\Big[\int_0^t (t-\tau)^{-\beta s}\tau^{-\alpha s}\,d\tau\Big]^{1/s}\;p_{s_2}(f)\\ &\leq K_{27}t^{(\delta+1/s-\beta-\alpha)}\;p_{s_2}(f) \leq K_{27}m(t)^{\varepsilon}\;p_{s_2}(f).\end{aligned}$$

In the case $t > 1$ we choose $k \in N$ such that $k < t \leq k+1$. We seperate the integral in the definition (52) into a sum

$$\int_0^t = \int_0^1 + \int_1^2 + \int_2^3 + \ldots + \int_{k-2}^{k-1} + \int_{k-1}^{t-1} + \int_{t-1}^t$$

and apply the Hölder inequality to each summand seperately. One gets

$$\begin{aligned}|g(t)| &\leq K_{26}p_{s_2}(f)\sum\Big[\int e^{-\lambda(t-\tau)s}(t-\tau)^{-\beta s}m(\tau)^{-\alpha s}\,d\tau\Big]^{1/s}\\ &\leq K_{26}p_{s_2}(f)\Big[\Big(\int_0^1\tau^{-\alpha s}d\tau\Big)^{1/s}e^{-\lambda(t-1)} + e^{-\lambda(t-2)} + e^{-\lambda(t-3)} + \ldots\\ &\qquad + e^{-\lambda(t-k+1)} + e^{-\lambda(t-t+1)} + \Big(\int_{t-1}^t(t-\tau)^{-\beta s}d\tau\Big)^{1/s}\Big]\\ &\leq K_{26}p_{s_2}(f)\Big[K(\alpha,\beta,s) + (e^{\lambda}-1)^{-1}\Big] \leq K_{28}m(t)^{\varepsilon}\;p_{s_2}(f).\end{aligned}$$

Note that the decomposition into a sum allows the restriction $t_2-t_1 \leq 1$ in the definition of $p_{s_2}(f)$, which avoids integrals over unbounded time intervals. Thus the Lemma is proved.

<u>Theorem 1</u> (Existence of mild solutions)

Let $p_0\in[1,\infty]$, $q_1,q_2\in[1,\infty]$ and $\gamma\in[1,\infty)$ satisfy

$$n/q_1 + 1/q_2 + (\gamma-1)n/p_0 < 1. \tag{54}$$

Assume for the initial data $u_0 \in L_{p_0}(\Omega)$ and for the nonlinear function F that (FO), the two-sided bound (F1) and Lipschitz condition (F3) hold with q_1,q_2,γ as stated above (see p.13).

Then the initial-boundary value problem (1)(1a)(1b) has a mild solution u on some time interval $[0,T]$. The solution u can be extended on a maximal time interval $[0,T_{max})$. Furtheron u satisfies (55)(56)(57)(58).

$$\lim_{t\to 0}\|u(t) - S(t)u_0\|_{p_0} = 0 \tag{55}$$

$$\lim_{t\to 0} \|u(t) - u_o\|_{p_o} = 0 \qquad \text{if assumption (A) p.20 holds.} \tag{56}$$

$$\sup\{\|u(t)\|_\infty \mid t \in [T_1,T_2]\} < \infty \qquad \text{for all } T_1,T_2 \in (0,T_{max}). \tag{57}$$

$$\lim_{t\to T_{max}} \|u(t)\|_{p_o} = \infty \qquad \text{if } T_{max} < \infty. \tag{58}$$

Assume furtheron that $1/q_1+(\gamma-1)/p_o \leq 1$ or $1/q_2+ n\gamma/p_o < 1$.

Let $p\in[p_o,\infty]$, $\delta\in[0,\infty)$ satisfy $n/p_o-n/p = \delta$.

Then the solution u satisfies

$$\sup\{m(t)^\delta \|u(t)\|_p \mid t \in (0,T_1]\} < \infty \qquad \text{for all } T_1 \in (0,T_{max}). \tag{59}$$

If the initial data satisfy $u_o \geqq 0$ and the function F satisfies the positivity assumption (F7), then the solution u satisfies $u \geqq 0$.

Remarks

(i) Uniqueness and maximality are specified in the Corollary p.54.

(ii) For initial data $u_o \in L_\infty(\Omega)$ Lemma 9 proves existence of a solution u under less restrictive assumptions on the nonlinearity F.

(iii) Formular (58) remains valid if the Lipschitz condition (F3) is replaced by the weaker local Lipschitz condition (F6). This is shown in Theorem 5(ii) p.76.

(iv) The awkward assumptions for (59) are real restrictions only for space dimension N = 1. They do not occur in Theorem 3 because of the primary a priori estimate (141).

Theorem 2 (Existence of mild solutions in the case of minimal regularity of the initial data)

Assume that the following "limit case" (L) occurs:

(L) $p_o,\gamma\in(1,\infty)$, $q_1\in[1,\infty]$ and $q_2\in(1,\infty]$ satisfy

$$n/q_1 + 1/q_2 + (\gamma-1)n/p_o = 1 \quad \text{(with equality sign!)}.$$

Assume for the initial data $u_o \in L_{p_o}(\Omega)$ and for the nonlinear function F that (F0), the two-sided bound (F1) and the Lipschitz condition (F3) hold with q_1,q_2,γ as stated above (see p.13).

Then the initial-boundary value problem (1)(1a)(1b) has a mild solution u on some time interval [0,T]. The solution u can be extended on a maximal time interval $[0,T_{max})$.

As in Theorem 1 the solution u satisfies (55)(56)(57). For initial data $u_o \geqq 0$ and nonlinearity F satisfying (F7) the solution u satisfies $u \geqq 0$. Formula (58) holds only under additional assumptions. Define the path Π and the set $k(\Pi)$ by

$$\Pi := \{u(t) \in L_{p_o} \mid t \in [0,T_{max})\} \quad \text{and} \tag{60}$$

$$k(\Pi) := \{u(t)/\|u(t)\|_{p_o} \mid t \in [0,T_{max}),\ \|u(t)\|_{p_o} \neq 0\}. \text{ Then} \tag{61}$$

$$\lim_{t \to T_{max}} \|u(t)\|_{p_o} = \infty \text{ if } T_{max} < \infty \text{ and the set } k(\Pi) \text{ is precompact in the space } L_{p_o}(\Omega). \tag{62}$$

The proof of Theorem 1 and 2 will be given in a sequence of Lemmas:

Lemma 7 uses a contraction argument in the Banach space $E_{p,\delta,T}$ (with a singularity at t = 0) to construct a solution u locally in time.

Lemma 8 considers the limit case (L) of minimal regularity of the initial data. A local solution u is constructed by a contraction argument in the Banach space $F_{w;p,\delta,T}$ defined below. Indeed, this space depends on the individual initial data u_o.

Lemma 9 improves the result of Lemma 7 for initial data $u_o \in L_\infty$.

Lemma 10 establishes positivity of solutions for positive initial data $u_o \in L_\infty$ and nonlinearities satisfying (F7).

Lemma 11 extends the result of Lemma 10 to initial data $u_o \in L_{p_o}$.

Lemma 12 constructs the finite chains of exponents $p_o < p_1 < \ldots < p_k$ $p_k = \infty$ and (δ_i) $i = 1\ldots k$ needed for the successive application of Lemma 7 and in the proof of Lemma 23.

Lemma 13 proves $u(t) \in L_\infty$ for all $t \in (0,T_{max})$.

Lemma 14 estimates $|u(t)|_\infty$ for small t.

Lemma 7

Let $p,p_o \in [1,\infty]$, $q_1,q_2 \in [1,\infty]$, $\gamma \in [1,\infty)$, $\delta \in [0,\infty)$ satisfy

$$\begin{aligned} n/q_1 + 1/q_2 + (\gamma-1)(\delta + n/p) + \varepsilon &< 1, \\ 1/q_1 + \gamma/p &\leqq 1, \\ 1/q_2 + \gamma\delta &< 1, \\ p_o \leqq p \quad \text{and} \quad n/p_o - n/p &= \delta. \end{aligned} \tag{63}$$

Take initial data $u_o \in L_{p_o}$ and assume for the nonlinear function F that (FO), the two-sided bound (F1) and the Lipschitz condition (F3) hold with q_1, q_2, γ from above.

Then for some $T \in (0,\infty)$ depending only on the exponents $p, p_o, q_1, q_2, \gamma, \delta, \varepsilon$ and the operator A_o as well as decreasingly on the norms $\|u_o\|_{p_o}, \|c\|_{q_1,q_2}$, the IBP (1)(1a)(1b) has a mild solution u on the time interval $[0,T]$. We have $u \in E_{p,\delta,T}$ (see p.11) and u is unique in this space. The solution u is independent of the choice of p and δ. Furtheron

$$\lim_{t\to 0} \|u(t) - S(t)u_o\|_{p_o} = 0. \tag{65}$$

<u>Proof</u>

Let $p, p_i \in [1,\infty]$ for $i = 1,2,3$. We use the Hölder inequalities

(H2) $\|u_1 u_2\|_p \leq \|u_1\|_{p_1} \|u_2\|_{p_2}$ for all $u_i \in L_{p_i}$ if $1/p_1 + 1/p_2 = 1/p$.

(H3) $\|u_1 u_2 u_3\|_p \leq \|u_1\|_{p_1} \|u_2\|_{p_2} \|u_3\|_{p_3}$ for all $u_i \in L_{p_i}$ if $1/p_1+1/p_2+1/p_3 = 1/p$.

To begin with, choose $T \in (0,\infty)$ and $u_1 \in E_{p,\delta,T}$ arbitrary. The number T will be diminished later. The proof uses a Picard-Lidelöf iteration. Define the sequence (u_m) by the iteration

$$u_{m+1}(t) = S(t)u_o + \int_0^t S(t-s)F(s,u_m(s))\,ds \quad \text{for all } t\in(0,T], m\in N \tag{66}$$

We show by induction on $m \in N$ that $u_m \in E_{p,\delta,T}$ is well defined for all $m \in N$. To this end, choose $s_1 \in [1,\infty]$ such that

$$1/q_1 + \gamma/p \leq 1/s_1 \quad \text{and} \quad n/s_1 + 1/q_2 + (\gamma-1)\delta + \varepsilon < 1 + n/p.$$

Assume that $u_m \in E_{p,\delta,T}$ is already proved. We estimate u_{m+1} defined by (66) using Hölder's inequality and the bound (F1):

$$m(t)^\delta \|u_{m+1}(t)\|_p \leq m(t)^\delta \|S(t)\|_{p,p_o} \|u_o\|_{p_o} +$$

$$+ m(t)^\delta \int_0^t \|S(t-s)\|_{p,s_1} m(s)^{-\gamma\delta} \|c(.,s)\|_{q_1} \left[m(s)^\delta \|1+|u_m|(s)\|_p\right]^\gamma ds.$$

The first term on the right-hand side can be estimated by Lemma 3 formula (30) since $0 \leq n/p_o - n/p = \delta$. To estimate the second term, we apply Lemma 6 with

$$f(s) = \|c(.,s)\|_{q_1}, \quad \alpha = \gamma\delta, \quad s_2 = q_2.$$

Note that assumption (51) of Lemma 6 is satisfied since

$$\begin{aligned} &n/s_1 + 1/q_2 < 1 + n/p, \quad && n/s_1 + 1/q_2 + \gamma\delta + \varepsilon < 1 + n/p + \delta, \\ &1/q_2 + \gamma\delta < 1, && 1/q_2 + \gamma\delta + \varepsilon \leq 1 + \delta. \end{aligned} \tag{67}$$

Note that $p_{s_2}(f) = \|c\|_{q_1,q_2}$. Hence we get by (53)

$$\begin{aligned} \|u_{m+1}\|_{p,\delta,T} &\leq K_{29}\|u_o\|_{p_o} + \\ &+ \sup_{0<t=T}\left\{m(t)^{\delta}\int_0^t \|S(t-s)\|_{p,s_1} m(s)^{-\gamma\delta} f(s)\, ds\right\} \|1+|u_m|\|^{\gamma}_{p,\delta,T} \\ &\leq K_{30}\left[\|u_o\|_{p_o} + m(T)^{\varepsilon}\|c\|_{q_1,q_2}\|1+|u_m|\|^{\gamma}_{p,\delta,T}\right]. \end{aligned} \tag{68}$$

This estimate proves $u_{m+1}(t) \in L_p$ for all $t \in (0,T]$ and $u_{m+1} \in E_{p,\delta,T}$.

Using the Lipschitz condition (F3) and (H3) one shows similarly

$$\|u_{m+1}-u_m\|_{p,\delta,T} \leq K_{31}m(T)^{\varepsilon}\|c\|_{q_1,q_2}\|1+|u_{m-1}|+|u_m|\|^{\gamma-1}_{p,\delta,T}\|u_m-u_{m-1}\|_{p,\delta,T}$$

$$\text{for all } m \in N\setminus\{1\}. \tag{69}$$

Let $U := \|u_1\|_{p,\delta,T} + 2\|u_2-u_1\|_{p,\delta,T}$ and choose $\tilde{T} \in (0,T]$ small enough such that

$$K_{31}m(\tilde{T})^{\varepsilon}\|c\|_{q_1,q_2}(1+2U)^{\gamma-1} \leq 1/2 \tag{70}$$

We restrict all functions u_m to the time interval $(0,\tilde{T}]$ and omit "~". By induction (69)(70) imply

$$\begin{aligned} \|u_{m+1}-u_m\|_{p,\delta,T} &\leq \frac{1}{2}\|u_m-u_{m-1}\|_{p,\delta,T} \quad && \text{for all } m \in N\setminus\{1\} \text{ and hence} \\ \|u_m\|_{p,\delta,T} &\leq U && \text{for all } m \in N. \end{aligned} \tag{71}$$

By (71) the sequence (u_m) is a Cauchy sequence in the Banach space $E_{p,\delta,T}$. Hence there exists a limit $u \in E_{p,\delta,T}$:

$$\lim_{m\to\infty} \|u_m - u\|_{p,\delta,T} = 0 \tag{72}$$

To estimate the nonlinearity F, we cancel $\|S(t-s)\|_{p,s_1}$ in (68). We get by the bound (F1) and Hölder's inequality

$$\|\!|F|\!\|_{1,1,T} \leq \int_0^T \|F(s,u(s))\|_{s_1}\, ds$$

$$\leq \int_0^T m(s)^{-\gamma\delta}\|c(.,s)\|_{q_1}\left[m(s)^{\delta}\|1+|u|(s)\|_p\right]^{\gamma} ds \tag{73}$$

$$\leq (K_{32} + T)\|c\|_{q_1,q_2}\|1+|u|\|^{\gamma}_{p,\delta,T}.$$

Using the Lipschitz condition (F3) and the Hölder inequality (H3) one shows by similar reasoning

$$\left\|\int_0^t S(t-s)[F(s,u_m(s))-F(s,u(s))]\, ds\right\|_1$$

$$\leq \int_0^t \|F(s,u_m(s))-F(s,u(s))\|_{s_1}\, ds \tag{74}$$

$$\leq \int_0^t m(s)^{-\gamma\delta}\|c(.,s)\|_{q_1}\, ds\ |1+|u_m|+|u||^{\gamma-1}_{p,\delta,t}|u_m-u|_{p,\delta,t}$$

$$\leq (K_{32} + t)\|c\|_{q_1,q_2}\|1+|u_m|+|u|\|^{\gamma-1}_{p,\delta,T}\|u_m - u\|_{p,\delta,T} \quad \text{for all } t\in(0,T].$$

To show that u is a mild solution of the IBP on the interval [0,T], we have to check that (a)(b)(c)(d) from Definition 1 p.29 hold.

(a) is satisfied since $u \in E_{p,\delta,T}$. (b) and (c) follow by estimate (73). To show that the integral equation (d) is satisfied, we take the limit $m \to \infty$ in the iteration (66). For all $t \in (0,T]$ the left-hand side of (66) converges in the $L_1(\Omega)$-norm by (72) and the right-hand side of (66) converges in the $L_1(\Omega)$-norm by (74) and (72). Hence taking the limit $m \to \infty$ in (66) yields the integral equation (45).

To show uniqueness, let $u_1, u_2 \in E_{p,\delta,T}$ be two mild solutions of the IBP. Similarly to (69) one gets

$$\|u_1-u_2\|_{p,\delta,T} \leq K_{31}m(T)^{\varepsilon}\|c\|_{q_1,q_2}\|1+|u_1|+|u_2|\|^{\gamma-1}_{p,\delta,T}\|u_1-u_2\|_{p,\delta,T} \tag{75}$$

Hence there exists $T_0 \in (0,T]$ such that $u_1(t) = u_2(t)$ for all $t \in [0,T_0]$. Uniqueness on the interval [0,T] follows by a continuation argument.

It remains to check that the solution u satisfies the initial condition. Assume that (64) holds. We choose $s_0\in[1,\infty]$ such that

$$1/q_1 + \gamma/p \leq 1/s_0 \quad\text{and}\quad n/s_0 + 1/q_2 + \gamma\delta + \varepsilon < 1 + n/p_0.$$

The computation ressembles (68). By bound (F1) and the Hölder inequality (H2) we estimate

$$\|u(t)-S(t)u_o\|_{p_o} \leq \left[\int_0^t \|S(t-s)\|_{p_o,s_o} m(s)^{-\gamma\delta} \|c(.,s)\|_{q_1} ds\right] \|1+|u|\|^{\gamma}_{p,\delta,t}.$$

Now apply Lemma 6 with $f(s) := |c(.,s)|_{q_1}$,

$$p := p_o\,,\ s_1 := s_o,\ s_2 := q_2,\ \alpha := \gamma\delta,\ \delta := 0,\ \varepsilon := \varepsilon.$$

Note that assumption (51) of Lemma 6 is satisfied since

$$n/s_o + 1/q_2 < 1 + n/p_o, \quad n/s_o + 1/q_2 + \gamma\delta + \varepsilon < 1 + n/p_o,$$
$$1/q_2 + \gamma\delta < 1, \qquad 1/q_2 + \gamma\delta + \varepsilon \leq 1.$$

Hence we get from (53) of Lemma 6 and the estimate above

$$\|u(t)-S(t)u_o\|_{p_o} \leq K_{33} m(t)^{\varepsilon} \|c\|_{q_1,q_2} \|1+|u|\|^{\gamma}_{p,\delta,t}, \tag{76}$$

which implies the assertion (65).
Thus the Lemma is proved.

<u>Lemma 8</u>

Let $p\in(1,\infty]$, $p_o\in(1,\infty)$, $q_1\in[1,\infty]$, $q_2\in(1,\infty]$, $\delta\in(0,\infty)$, $\gamma\in(1,\infty)$ satisfy

$$\begin{aligned} n/q_1 + 1/q_2 + (\gamma-1)(\delta + n/p) &= 1 \quad \text{(with equality sign!)},\\ 1/q_1 + \gamma/p &\leq 1,\\ 1/q_2 + \gamma\delta &< 1,\\ n/p_o - n/p &= \delta. \end{aligned} \tag{77}$$

Take initial data $u_o \in L_{p_o}$ and assume for the nonlinear function F that (F0), the two-sided bound (F1) and the Lipschitz condition (F3) hold with q_1,q_2,γ from above.

Then for some $T \in (0,\infty)$ depending on the exponents p,p_o,q_1,q_2,γ, and the operator A_o as well as on the initial data u_o (not only on $\|u_o\|_{p_o}$!) and the norm $\|c\|_{q_1,q_2}$, the IBP (1)(1a)(1b) has a mild solution on the time interval $[0,T]$. If the set $\Pi \subset L_{p_o}$ is bounded and has the property (K) p.27 (i.e. $k(\Pi) = \{u/\|u\|_{p_o} \mid u\in L_{p_o},\ u\neq 0\}$ is precompact in L_{p_o}), then the interval $[0,T]$ is independent of $u_o \in \Pi$. Furtheron the solution $u \in F_{W;p,\delta,T} \subset E_{p,\delta,T}$ is unique only in the subspace $F_{W;p,\delta,T}$ defined below. The initial data u_o is assumed in the sense (65).

Proof

Estimate (41) of Lemma 4 is applied with $p := p_0$, $q := p$ and $\Pi := \{u_0\} \subset L_{p_0}$. Let $g = g(t)$ and $W = W(t)$ be the functions constructed in Lemma 4 and recall that $W(t)^{-\delta} = g(t)m(t)^{-\delta}$.

Let $F_{W;p,\delta,T} \subset E_{p,\delta,T}$ be the Banach space of all functions $u \in E_{p,\delta,T}$

such that $\|u\|_{W;p,\delta,T} := \sup\{W(t)^{\delta}\|u(.,t)\|_p \mid t \in (0,T]\} < \infty.$ (78)

By Lemma 4(c) we know $m(t) \leq W(t)$. Hence $\|u\|_{W;p,\delta,T} \geq \|u\|_{p,\delta,T}$.

The existence proof is based on the iteration (66), which takes place in the Banach space $F_{W;p,\delta,T}$. To begin with, choose $T \in (0,\infty)$ and $u_1 \in F_{W;p,\delta,T}$ arbitrary. We prove by induction on m that $u_m \in F_{W;p,\delta,T}$ for all $m \in N$. Choose $s_1 \in [1,\infty]$ such that (equality)

$$n(1/q_1 + \gamma/p) = n/s_1 = 1 + n/p - 1/q_2 - (\gamma-1)\delta.$$

Assume that $u_m \in F_{W;p,\delta,T}$ is already proved. We estimate u_{m+1} using Hölder's inequality and the bound (F1):

$$\|u_{m+1}\|_{W;p,\delta,T} \leq K_{34}\|u_0\|_{p_0} + \varphi(T)\|1+|u_m|\|^{\gamma}_{W;p,\delta,T} \quad \text{where} \tag{79}$$

$$\varphi(T) = \sup\left\{W(t)^{\delta}\int_0^t \|S(t-s)\|_{p,s_1} W(s)^{-\gamma\delta}\|c(.,s)\|_{q_1}\, ds \,\middle|\, t\in(0,T]\right\}$$

$$= g(T)^{\gamma-1}\sup\left\{m(t)^{\delta}\int_0^t \|S(t-s)\|_{p,s_1} m(s)^{-\gamma\delta}\|c(.,s)\|_{q_1}\, ds \,\middle|\, t\in(0,T]\right\}$$

To estimate the last term we apply Lemma 6 with $f(s) := \|c(.,s)\|_{q_1}$, $p := p$, $s_1 := s_1$, $s_2 := q_2$, $\alpha := \gamma\delta$, $\delta := \delta$, $\varepsilon := 0$. Hence

$$\varphi(T) \leq K_{34}g(T)^{\gamma-1}\|c\|_{q_1,q_2} \quad \text{and} \quad \lim_{T\to 0}\varphi(T) = 0. \tag{80}$$

Estimates (79)(80) prove $u_{m+1}(t) \in L_p$ for all $t\in(0,T]$ and $u_{m+1} \in F_{W;p,\delta,T}$. By the Lipschitz condition (F3) and Hölder's inequality (H3) one gets

$$\|u_{m+1}-u_m\|_{W;p,\delta,T} \leq \varphi(T)\|1+|u_{m-1}|+|u_m|\|^{\gamma-1}_{W;p,\delta,T}\|u_m-u_{m-1}\|_{W;p,\delta,T}$$

$$\text{for all } m \in N\setminus\{1\}. \tag{81}$$

Now we argue as in the proof of Lemma 7.

Let $U := \|u_1\|_{W;p,\delta,T} + 2\|u_1 - u_2\|_{W;p,\delta,T}$ and choose $T \in (0,\tilde{T}]$ such that

$$\varphi(\tilde{T})(1 + 2U)^{\gamma-1} \leq 1/2. \tag{82}$$

We restrict all functions u_m to the time interval $(0,\tilde{T}]$ and omit "$\tilde{}$". By induction (81)(82) imply

$$\|u_{m+1} - u_m\|_{W;p,\delta,T} \leq \frac{1}{2}\|u_m - u_{m-1}\|_{W;p,\delta,T} \quad \text{for all } m \in N\setminus\{1\} \text{ and hence}$$

$$\|u_m\|_{W;p,\delta,T} \leq U \quad \text{for all } m \in N.$$

Hence the sequence (u_m) has a limit $u \in F_{W;p,\delta,T}$:

$$\lim_{m\to\infty} \|u_m - u\|_{W;p,\delta,T} = 0 \tag{83}$$

Estimates (73) and (74) concerning the nonlinearity F are proved exactly as in Lemma 7. Hence the function u is a mild solution of the IBP on the time interval $[0,T]$.

It remains to check that the solution u satisfies the initial condition. We estimate using the bound (F1) and Hölder's inequality

$$\|u(t) - S(t)u_0\|_{p_0} \leq \left[\int_0^t \|S(t-s)\|_{p_0,s_1} W(s)^{-\gamma\delta} \|c(.,s)\|_{q_1} ds\right] \|1+|u|\|^{\gamma}_{W;p,\delta,t}.$$

To estimate the left-hand side we apply Lemma 6 with $f(s) := |c(.,s)|_{q_1}$,

$$p := p_0,\ s_1 := s_1,\ s_2 := q_2,\ \alpha := \gamma\delta,\ \delta := 0,\ \varepsilon := 0.$$

Note that assumption (51) of Lemma 6 is satisfied since

$$n/s_1 + 1/q_2 < 1 + n/p_0, \quad n/s_1 + 1/q_2 + \gamma\delta \leq 1 + n/p_0,$$
$$1/q_2 + \gamma\delta < 1.$$

Hence we get

$$\|u(t) - S(t)u_0\|_{p_0} \leq K_{35} g(t)^{\gamma} \|c\|_{q_1,q_2} \|1+|u|\|^{\gamma}_{W;p,\delta,t}. \tag{84}$$

Now (80) and (84) imply the assertion (65).

If the set of initial data $\Pi \subset L_{p_0}$ is bounded and has property (K), the function $W = W(t)$ in Lemma 4 as well as the numbers $U, T \in (0,\infty)$ from above can be chosen independent of $u_0 \in \Pi$. Thus the Lemma is proved.

Lemma 9

Let $q_1, q_2 \in [1,\infty]$ and $\varepsilon \in (0,1)$ satisfy

$$n/q_1 + 1/q_2 + \varepsilon < 1. \tag{85}$$

Take initial data $u_o \in L_\infty$ and assume for the nonlinear function F that (F0), the local two-sided bound (F4) and the local Lipschitz condition (F6) hold with q_1, q_2 from above.

Then for some $T \in (0,\infty)$ depending only on the exponents q_1, q_2, the operator A_o, the function Γ occuring in (F4)(F6) and decreasingly on $\|u_o\|_\infty$, $\|c\|_{q_1,q_2}$, the IBP (1)(1a)(1b) has a mild solution u on the time interval $[0,T]$. We have $u \in E_{\infty,0,T}$ and u is unique in this space. Furtheron

$$\lim_{t\to 0} \|u(t) - S(t)u_o\|_\infty = 0. \tag{86}$$

Proof

The proof is based on the iteration (66) which takes place in the Banach space $E_{\infty,0,T}$. To begin with, choose $T \in (0,\infty)$ and $u_1 \in E_{\infty,0,T}$ arbitrary. We show by induction on m that $u_m \in E_{\infty,0,T}$ for all $m \in N$. Assume that $u_m \in E_{\infty,0,T}$ is already shown. We estimate u_{m+1} by the bound (F4):

$$\|u_{m+1}(t)\| \le K_{36}\|u_o\|_\infty + \Gamma(\|u_m\|_{\infty,0,T}) \int_0^t \|S(t-s)\|_{\infty,q_1} \|c(.,s)\|_{q_1}\, ds.$$

Now apply Lemma 6 with $f(s) := \|c(.,s)\|_{q_1}$,

$p := \infty$, $s_1 := q_1$, $s_2 := q_2$, $\alpha := 0$, $\delta := 0$, $\varepsilon := \varepsilon$.

Assumption (51) of Lemma 6 is satisfied by (85). Hence we get

$$\|u_{m+1}\|_{\infty,0,T} \le K_{37}[\|u_o\|_\infty + m(T)^\varepsilon \|c\|_{q_1,q_2} \Gamma(\|u_m\|_{\infty,0,T})], \tag{87}$$

which proves $u_{m+1} \in E_{\infty,0,T}$.

Using the local Lipschitz condition (F6) and the Hölder inequality (H3) one shows by similar arguments

$$\|u_{m+1} - u_m\|_{\infty,0,T} \le K_{38} m(T)^\varepsilon \|c\|_{q_1,q_2} \Gamma(\|u_{m-1}\| + \|u_m\|_{\infty,0,T}) \|u_m - u_{m-1}\|_{\infty,0,T}$$

$$\text{for all } m \in N\setminus\{1\}. \tag{88}$$

Let $U := \|u_1\|_{\infty,0,T} + 2\|u_2 - u_1\|_{\infty,0,T}$ and choose $\tilde{T} \in (0,T]$ such that

$$2K_{38}m(\tilde{T})^{\varepsilon}\|c\|_{q_1,q_2}\Gamma(2U) \le 1/2. \quad (2K_{38} \text{ is used in Lemma 10}) \tag{89}$$

We restrict all functions u_m to the time interval $(0,\tilde{T}]$ and omit "$\tilde{}$". By induction (88)(89) imply

$$\|u_{m+1}-u_m\|_{\infty,0,T} \le \frac{1}{2}\|u_m-u_{m-1}\|_{\infty,0,T} \quad \text{for all } m \in N\setminus\{1\} \text{ and hence}$$
$$\|u_m\|_{\infty,0,T} \le U \quad \text{for all } m \in N. \tag{90}$$

Hence (u_m) is a Cauchy sequence in the Banach space $E_{\infty,0,T}$ and there exists a limit $u \in E_{\infty,0,T}$:

$$\lim_{m\to\infty} \|u_m - u\|_{\infty,0,T} = 0. \tag{91}$$

We consider the nonlinearity. Let $F = F(x,t,u(x,t))$, $F_m = F(x,t,u_m(x,t))$. The local Lipschitz condition (F6) implies

$$\lim_{m\to\infty} \|F_m - F\|_{\infty,0,T} = 0.$$

Hence we can take the limit $m \to \infty$ of the iteration (66) in the norm of $E_{\infty,0,T}$. We see that the function u satisfies the integral equation (45) and is indeed a mild solution of the IBP on the interval $[0,T]$.

Let $u_1,u_2 \in E_{\infty,0,T}$ be two solutions of the IBP. Estimate (88) shows that there exists $T_0\in(0,T]$ such that $u_1(t) = u_2(t)$ for all $t \in (0,T_0]$. Now uniqueness on the interval $[0,T]$ follows by a continuation argument.

It remains to check that the solution u satisfies the initial condition. We estimate u using the bound (F4)

$$\|u(t)-S(t)u_0\|_\infty \le \Gamma(\|u\|_{\infty,0,T})\int_0^t \|S(t-s)\|_{\infty,q_1}\|c(.,s)\|_{q_1}\,ds.$$

Hence the arguments leading to (87) show

$$\|u(t)-S(t)u_0\|_\infty \le K_{37}m(t)^{\varepsilon}\|c\|_{q_1,q_2}\Gamma(\|u\|_{\infty,0,T}), \tag{92}$$

which implies (86). Thus the Lemma is proved.

Lemma 10

Make the assumption of Lemma 9 and furtheron assume $u_0 \geqq 0$ and the positivity assumption (F7) for the nonlinearity F. Then the solution u of the IBP constructed in Lemma 9 satisfies $u \geqq 0$.

Proof

Let $T,U \in (0,\infty)$ be the numbers choosen in the proof of Lemma 9 p.42 . We have to smooth the nonlinearity F. Choose a sequence of smoothing kernels $\varepsilon_k \in C(R^N \times R)$ for all $k \in N$ with the following properties:

$$\varepsilon_k(y,s) \geqq 0 \quad \text{for all } (y,s) \in R^N \times R,$$

$$\varepsilon_k(y,s) = 0 \quad \text{for all } (y,s) \in R^N \times R \text{ such that } |y|+|s| > 2^{-k},$$

$$\int_{R^{N+1}} \varepsilon_k(y,s)\, dyds = 1.$$

We define the smoothing operators E_k: $f \in L_1(\Omega\times[0,T]) \to E_k f \in L_1(\Omega\times[0,T])$ by

$$(E_k f)(x,t) = \int_\Omega \int_0^T \varepsilon_k(x-y,t-s) f(y,s)\, dyds \quad \text{for all } (x,t) \in \bar{\Omega}\times[0,T].$$

We need the following properties of the positive, linear operators E_k:

(E1) $|||E_k f|||_{\infty,\infty,T} \leq |\Omega| T \max\limits_{(y,s)} \varepsilon_k(y,s)\, |||f|||_{1,1,T}$ for all $f \in L_{1,1,T}$.

(E2) For all $f \in C(\bar{\Omega}\times[0,T])$ and all compact sets $\Sigma \subset \mathrm{int}[\Omega\times(0,T)]$ $\lim\limits_{k\to\infty} \sup \{ |[(E_k f) - f](x,t)| \ (x,t) \in \Sigma\} = 0$.

(E3) $|||E_k f|||_{q_1,q_2,T} \leq 2^{1/q_2} |||f|||_{q_1,q_2,T}$ for all $f \in L_{q_1,q_2,T}$.

(E4) If $q_1,q_2 \in [1,\infty)$ and $f \in L_{q_1,q_2,T}$, then $\lim\limits_{k\to\infty} |||(E_k f) - f|||_{q_1,q_2,T} = 0$.

(E1) is obvious. (E2) is an easy consequence of the uniform continuity of f. To show (E3) we use Young's inequality (see Reed and Simon[52]p.29)

$$\|f*g\|_p \leq \|f\|_1 \|g\|_p \quad \text{for all } f \in L_1(R^N),\ g \in L_p(R^N), \text{ where}$$

$$(f*g)(x) = \int_{R^N} f(x-y) g(y)\, dy.$$

Let $f \in L_{q_1,q_2,T}$ and define the function $\varepsilon_k * f \in L_1(\Omega\times[0,T]^2)$ by

$$(\varepsilon_k * f)(x,t,s) = \int_\Omega \varepsilon_k(x-y,t-s) f(y,s)\, dy \quad \text{for all } (x,t,s) \in \bar{\Omega}\times[0,T]^2.$$

Young's inequality implies

$$\|(\varepsilon_k * f)(.,t,s)\|_{q_1} \leq \left(\int_{R^N} \varepsilon_k(x,t-s)\,dx\right) \|f(.,s)\|_{q_1} \quad \text{for all } (t,s)\in[0,T]^2.$$

Since by definition

$$(E_k f)(x,t) = \int_0^T (\varepsilon_k * f)(x,t,s)\,ds \quad \text{for all } (x,t) \in \Omega\times[0,T],$$

we can estimate

$$\|(E_k f)(.,t)\|_{q_1} \leq \int_0^T\left(\int_{R^N} \varepsilon_k(x,t-s)\,dx\right)\|f(.,s)\|_{q_1} ds \quad \text{for all } t\in[0,T]. \tag{93}$$

Let $t_1, t_2 \in [0,T]$ such that $0 \leq t_2 - t_1 \leq 1$ and $s_1 = \max\{0, t_1 - 1/2\}$, $s_2 = \min\{T, t_2 + 1/2\}$. Applying Young's inequality once more to the time convolution in (93) yields

$$\left[\int_{t_1}^{t_2} \|E_k f(.,t)\|_{q_1}^{q_2}\right]^{1/q_2} \leq \int_{R^{N+1}} \varepsilon_k(x,t)\,dxdt \left[\int_{s_1}^{s_2} \|f(.,s)\|_{q_1}^{q_2} ds\right]^{1/q_2}$$

$$\leq 2^{1/q_2} |||f|||_{q_1,q_2,T} \quad \text{and hence (E3).}$$

To prove (E4) choose a sequence f_l in the space $C(\overline{\Omega}\times[0,T])$ such that $\lim_{l\to\infty} |||f_l - f|||_{q_1,q_2,T} = 0$. (E3) and the triangle inequality imply

$$|||E_k f - f|||_{q_1,q_2,T} \leq |||E_k f_l - f_l|||_{q_1,q_2,T} + 3|||f_l - f|||_{q_1,q_2,T}$$

Let $\varepsilon > 0$ be given. Then for some $l \in N$ the second term on the left-hand side become less than ε. Now keep l fixed. Then by (E2) the first term becomes less than ε for $k \in N$ large enough. Hence (E4) follows.

Let $c \in L_{q_1,q_2}$ be the function occuring in the local bound (F4) and the local Lipschitz condition (F6). Since $L_{\infty,q_2} \subset L_{q_1,q_2}$ and $L_{q_1,\infty} \subset L_{q_1,q_2}$ for all $q_1,q_2\in[1,\infty]$, we may assume without any restriction $q_1,q_2\in[1,\infty)$. (Otherwise the topology of L_{q_1,q_2} is so strong that (E4) does not hold). We define the smoothed functions F_k and c_k by

$$F_k(x,t,u) = \int_\Omega\int_0^T \varepsilon_k(x-y,t-s)F(y,s,u)\,dyds \quad \text{for all } (x,t,u)\in\overline{\Omega}\times[0,T]\times R,$$

$$c_k(x,t) = \int_\Omega\int_0^T \varepsilon_k(x-y,t-s)c(y,s)\,dyds \quad \text{for all } (x,t) \in \overline{\Omega}\times[0,T].$$

Then (F4)(F6)(F7) imply for all $(x,t) \in \bar{\Omega}\times[0,T]$:

$$|F_k(x,t,u)| \leq c_k(x,t)\Gamma(|u|) \qquad \text{for all } u \in R, \tag{94}$$

$$|F_k(x,t,u)-F_k(x,t,v)| \leq c_k(x,t)\Gamma(|u|+|v|)|u - v| \text{ for all } u,v\in R, \tag{95}$$

$$F_k(x,t,0) \geqq 0. \tag{96}$$

Hence (E3)(E4) and (94) imply

$$\lim_{k\to\infty} |||(F_k - F)(.,.,u)|||_{q_1,q_2,T} = 0 \quad \text{for all } u \in R. \tag{97}$$

By (E3) and (94)(95) the functions F_k: $u\in[-U,U] \to F_k(.,.,u)\in L_{q_1,q_2,T}$ are uniformly bounded and equicontinuous for all $k \in N$.
Hence (97) and the Theorem of Ascoli imply

$$\lim_{k\to\infty} \sup \{|||(F_k - F)(.,.,u)|||_{q_1,q_2,T} \mid u\in[-U,U]\} = 0 \quad \text{for all } U \in R. \tag{98}$$

Since by (E3) we have $\sup_{k\in\bar{N}} |||c_k|||_{q_1,q_2,T} \leq 2\|c\|_{q_1,q_2}$, (94)(95) and the proof of Lemma 9 imply $\sup_{k\in\bar{N}} \|u_k\|_{\infty,0,T} \leq U < \infty.$ (99)

Define for all $k \in N$

$$a_k = |||c_k|||_{\infty,\infty,T}\,\Gamma(U),$$

$$G_k(x,t,u) = F_k(x,t,u) + a_k u \quad \text{for all } (x,t,u) \in \bar{\Omega}\times[0,T]\times R,$$

$$v_k(x,t) = u_k(x,t)e^{a_k t} \qquad \text{for all } (x,t) \in \bar{\Omega}\times[0,T].$$

By (94)(95)(96)(99) we get

$$G_k(x,t,u_k(x,t)) \geqq 2a_k \min\{0,u_k(x,t)\} \quad \text{for all } (x,t) \in \Omega\times[0,T].$$

The integral equation (45) implies

$$v_k(t) = S(t)u_0 + \int_0^t S(t-s)\, e^{a_k s}G_k(s,u_k(s))\, ds \quad \text{for all } t\in[0,T]. \tag{100}$$

Define the function $w_k = \max\{0,-v_k\}$. Since $u_0 \geqq 0$ by assumption and the semigroup $S(t)$ is positive by Lemma 2(i) we get

$$w_k(t) \leq 2a_k\int_0^t S(t-s)w_k(s)\, ds \quad \text{for all } t\in[0,T] \text{ and hence}$$

$$\| w_k(t) \|_\infty \leq 2a_k K_{39} \int_0^t \| w_k(s) \|_\infty ds \quad \text{for all } t \in [0,T].$$

Hence the Lemma of Gronwall implies $w_k \equiv 0$ and thus $v_k \geqq 0$, $u_k \geqq 0$. for all $k \in N$.

We take the limit $k \to \infty$ in the approximation process. The integral equation (45) implies

$$(u_k - u)(t) = \int_0^t S(t-s)[F(s,u(s))-F(s,u_k(s))]\, ds + \\ + \int_0^t S(t-s)[F(s,u_k(s))-F_k(s,u_k(s))]\, ds. \tag{101}$$

By the same methods as in Lemma 9 this implies

$$\| u_k - u \|_{\infty,0,T} \leq K_{38} \Gamma(2U) m(T)^\varepsilon \| c \|_{q_1,q_2} \| u_k - u \|_{\infty,0,T} + \\ + K_{40} \sup\{ ||| (F_k - F)(.,.,u) |||_{q_1,q_2,T} \mid u \in [-U,U] \} \tag{102}$$

By (89) the number T is small enough such that we get

$$\| u_k - u \|_{\infty,0,T} \leq 2K_{40} \sup \{ ||| (F_k - F)(.,.,u) |||_{q_1,q_2,T} \mid u \in [-U,U] \} \tag{103}$$

Hence (98) implies

$$\lim_{k\to\infty} \| u_k - u \|_{\infty,0,T} = 0. \tag{104}$$

Now $u_k \geqq 0$ for all $k \in N$ implies $u \geqq 0$. Thus the Lemma is proved.

Lemma 11

Make the assumptions of Lemma 7 or 8 and furtheron let the initial data satisfy $u_o \geqq 0$ and let the nonlinearity F satisfy the positivity assumption (F7). Then the mild solution u of the IBP constructed in Lemma 7 or 8, respectively satisfies $u \geqq 0$.

Proof

The case $p_o = \infty$ is considered in Lemma 10. First make the assumptions of Lemma 7. Let $p_o \in [1,\infty)$, $u_o \in L_{p_o}$, $u_o \geqq 0$ be given. We choose a sequence (u_{om}) in the space L_∞ such that

$$\lim_{m\to\infty} \| u_{om} - u_o \|_{p_o} = 0, \quad u_{om} \geqq 0 \text{ and } \| u_{om} \|_{p_o} \leq \| u_o \|_{p_o} \text{ for } m \in N. \tag{105}$$

By Lemma 7 there exist $T,U \in (0,\infty)$ and mild solutions $u_m \in E_{p,\delta,T}$ of the IBP to initial data u_{om} on the time interval $[0,T]$ such that

$$\sup\{\|u_m\|_{p,\delta,T} \mid m \in N\} \leq U. \tag{106}$$

(The mere existence of mild solutions could also be infered from Lemma 9, but not the uniform properties expressed by the numbers T and U). Similarly to (68)(69) on gets the estimate

$$\begin{aligned}\|u_m - u\|_{p,\delta,T} &\leq K_{29}\|u_{om} - u_o\|_{p_o} \\ &\quad + K_{31}m(T)^{\varepsilon}\|c\|_{q_1,q_2}\|1+|u|+|u_m|\|^{\gamma-1}_{p,\delta,T}\|u_m - u\|_{p,\delta,T}.\end{aligned} \tag{107}$$

Hence (106) and (70)(which guarantees that T is small) imply

$$\|u_m - u\|_{p,\delta,T} \leq 2K_{29}\|u_{om} - u_o\|_{p_o}. \tag{108}$$

By Lemma 10 we get $u_m \geqq 0$ for all $m \in N$. Hence (105)(108) imply $u \geqq 0$.

Now let $p_o \in (1,\infty)$, $u_o \in L_{p_o} \setminus \{0\}$, $u_o \geqq 0$ and make the assumptions of Lemma 8. The approximating sequence u_{om} in L_∞ can be chosen such that

$$\Pi := \{u_o, u_{om} \mid m \in N\} \subset L_{p_o} \setminus \{0\}.$$

We begin the proof of Lemma 8 by applying estimate (41) of Lemma 4 with $p := p_o$, $q := p$ and Π from above. As shown in the proof of Lemma 8, there exist $T,U \in (0,\infty)$ and mild solutions $u_m \in F_{W;p,\delta,T}$ of the IBP to initial data u_{om} on the time interval $[0,T]$ such that

$$\sup\{\|u_m\|_{W;p,\delta,T} \mid m \in N\} \leq U. \tag{109}$$

Similarly to (79)(81) one gets the estimate

$$\begin{aligned}\|u_m - u\|_{W;p,\delta,T} &\leq K_{34}\|u_{om} - u_o\|_{p_o} \\ &\quad + \varphi(T)\|1+|u|+|u_m|\|_{W;p,\delta,T}\|u_m - u\|_{W;p,\delta,T}.\end{aligned} \tag{110}$$

Hence (109) and (82)(which guarantees that T is small) imply

$$\|u_m - u\|_{W;p,\delta,T} \leq 2K_{34}\|u_{om} - u_o\|_{p_o} \tag{111}$$

By Lemma 10 we get $u_m \geqq 0$ for all $m \in N$. Hence (105)(111) imply $u \geqq 0$. Thus the Lemma is proved.

Lemma 12

Consider the following two cases (i) and (ii)(limit case)

(i) Let $q_1, q_2 \in [1,\infty]$, $p_o, \gamma \in [1,\infty)$ satisfy

$$n/q_1 + 1/q_2 + (\gamma-1)n/p_o < 1. \tag{54}$$

(ii) Let $q_1 \in [1,\infty]$, $q_2 \in (1,\infty]$, $p_o, \gamma \in (1,\infty)$ satisfy

$$n/q_1 + 1/q_2 + (\gamma-1)n/p_o = 1 \quad \text{(with equality sign).} \tag{L}$$

Then there exist $\varepsilon \in (0,1)$, $k \in N$ and finite sequences (p_i) and (δ_i) such that

$$p_o < p_1 < p_2 < \ldots < p_k = \infty,$$

$$\delta_i \in (0,1) \quad \text{for } i = 1\ldots k$$

and (112)(113)(114)(115) hold.

$$n/q_1 + 1/q_2 + (\gamma-1)(\delta_i + n/p_i) + \varepsilon < 1 \quad \text{for } i = 1\ldots k, \; i \neq 1 \text{ in case (ii).} \tag{112}$$

$$1/q_1 + \gamma/p_i \leq 1 \quad \text{for } i = 1\ldots k \tag{113}$$

$$1/q_2 + \gamma\delta_i < 1 \quad \text{for } i = 1\ldots k \tag{114}$$

$$n/p_{i-1} - n/p_i = \delta_i \quad \text{for } i = 1\ldots k \tag{115}$$

Proof

The important point is to choose $p_1 \in [1,\infty]$ such that (113)(114) hold with δ_1 given by (115). These conditions for p_1 can be written more clearly:

$$\begin{aligned} 0 \leq \gamma n/p_1 &< \gamma n/p_o, \\ \gamma n/p_1 &\leq n - n/q_1, \\ 1/q_2 + \gamma n/p_o - 1 &< \gamma n/p_1. \end{aligned} \tag{116}$$

There exists some $p_1 \in [1,\infty]$ satisfying (116) if and only if (117) holds:

$$\begin{aligned} &0 < 1/p_o, \quad 1/q_1 \leq 1, \quad 1/q_2 < 1, \\ &n/q_1 + 1/q_2 + (\gamma-1)n/p_o < 1 + n(1 - 1/p_o). \end{aligned} \tag{117}$$

Under the assumptions of the Lemma (117) and hence (116) holds. Thus there exist p_1 and δ_1 satisfying (112)(113)(114)(115) for i = 1. Now choose $p_2 < p_3 < \dots < p_k = \infty$ and (δ_i) for $i = 2\dots k$ such that

$$\delta_i = n/p_{i-1} - n/p_i \leq \delta_1 \quad \text{for } i = 2\dots k.$$

Then all assertions of the Lemma are satisfied.

Definition 2

Let $T_1 \in [0,\infty)$, $T_2 \in (0,\infty)$, $r \in [1,\infty]$.

By a L_r-mild solution of the IBP for initial data $v \in L_1(\Omega)$ on the time interval $[T_1, T_1+T_2]$ we mean a measurable function u,

$u: (x,t) \in \Omega\times(T_1,T_1+T_2] \to u(x,t) \in R$ such that $u(t) = u(.,t)$ satisfies

(a) $u(t) \in L_r(\Omega)$ for all $t \in (T_1, T_1+T_2]$,

(b) $F(t,u(t)) \in L_1(\Omega)$ for almost all $t \in (T_1,T_1+T_2)$,

(c) $\int_{T_1}^{T_1+T_2} \|F(t,u(t))\|_1 \, dt < \infty$,

(d) $u(T_1+t) = S(t)v + \int_0^t S(t-s)F(T_1+s,u(T_1+s)) \, ds$ for all $t \in (0,T_2]$.

where the integral is an absolutely converging Bochner-integral in the space $L_1(\Omega)$.

A L_1-mild solution is simply called a mild solution. For $T_1 = 0$, $r = 1$ we are back in the case of Defintion 1, p.29.

Lemma 13

Make the assumptions of Theorem 1 or 2. Then for some $T \in (0,\infty)$ the IBP has a mild solution u for initial data u_o on the time interval $[0,T]$. This solution assumes the initial data u_o in the sense of (55). Furtheron

$$\sup \{\|u(t)\|_\infty | t \in [T_o,T]\} < \infty \quad \text{for all } T_o \in (0,T]. \tag{118}$$

Proof

The case $p_o = \infty$ is treated in Lemma 9. Assume $p_o < \infty$ and let $p_o < p_1 < \dots < p_k = \infty$ and (δ_i) $i = 1\dots k$ be the finite sequences constructed in Lemma 12 (i) or (ii) in the case of Theorem 1 or 2, respectively. We apply Lemma 7 or 8, respectively, with $p_o := p_o$, $p := p_1$, $\delta := \delta_1$

and initial data u_o. By Lemma 7 or 8 there exists $T \in (0,\infty)$ and a mild solution u of the IBP on the interval [0,T] satisfying (65).
We show by induction on i that this solution u satisfies for i = 1...k

$$\sup \{\|u(t)\|_{p_i} \mid t\in[T_o,T]\} < \infty \text{ for all } T_o \in (0,T]. \tag{119_i}$$

Since $u \in E_{p_1,\delta_1,T}$ is proved in Lemma 7 or 8, (119_i) holds for i = 1. Assume that (119_{i-1}) is already shown for some i. We have to prove (119_i). Let $T_o \in (0,T]$ be arbitrary. For all $s \in [T_o,T]$ we apply Lemma 7 with $p_o := p_{i-1}$, $p := p_i$, $\delta := \delta_i$ and $u_o := u(s)$ as initial data. Hence there exists $T_i \in (0,\infty)$ (which is independent of s by (119_{i-1})) and mild solutions u_i^s for initial data u(s) on the time intervals $[s,s+T_i]$ for all $s \in [T_o,T]$. Furtheron $u_i^s(.,s+.) \in E_{p_i,\delta_i,T_i}$ is unique in this space. Hence $u_i^s(t) = u(t)$ for all $t \in (s,s+T_i]\cap(0,T]$. Lemma 7 yields furtheron

$$\sup \{\|u_i^s\|_{p_i,\delta_i,T_i} \mid s\in[T_o,T]\} < \infty.$$

Since $T_o \in (0,T]$ is chosen arbitrary this implies (119_i). By induction we get (119_k) which is the assertion of the Lemma.

<u>Lemma 14</u>
Make the assumptions of Theorem 1 and let u be the mild solution of the IBP on the interval [0,T] given in Lemma 13. Furtheron let (120a) or (120b) hold.

$$1/q_2 + \gamma n/p_o < 1 , \qquad 1/q_1 + (\gamma-1)/p_o \leq 1 \tag{120a),(120b}$$

Then there exists $M\in(0,\infty)$ such that

$$\|u\|_{p,\delta,T} \leq M \quad \text{for all } p\in[p_o,\infty],\ \delta = n/p_o-n/p. \tag{121}$$

<u>Proof</u>
Assume $p_o\in[1,\infty)$, since the case $p_o = \infty$ is clear. Let $V\in(0,\infty)$ and $B(V) = \{v \in L_{p_o}(\Omega) \mid \|v\|_{p_o} \leq V\}$. By the proof of Lemma 7 p.36 there exist $T,U\in(0,\infty)$ independent of $u_o\in B(V)$ such that with p,δ as in Lemma 7 we get $\|u\|_{p,\delta,T} \leq U$ for the mild solution u of the IBP for initial data u_o on the interval [0,T]. Furtheron (76) and (30) imply the estimate

$$\|u(t)\|_{p_o} \leq \|S(t)u_o\|_{p_o} + K_{33}\|c\|_{q_1,q_2}(1+\|u\|_{p,\delta,t})^{\gamma} \quad \text{and hence}$$

$$\|u(t)\|_{p_o} \leq K_{13}V + K_{33}\|c\|_{q_1,q_2}(1+U)^{\gamma} \quad \text{for all } t\in(0,T], \text{ all solutions } u \text{ for initial data } u_o\in B(V). \tag{122}$$

Consider the case that (120b) is assumed. We use part of Theorem 3 below (which will indeed be proved without using the present Lemma). Let $r_1 := p_o$, $r_2 := \infty$ in Theorem 3. After restricting attention to initial data $u_o \in B(V)$, (122) is an a priori estimate of the type (141) assumed in Theorem 3. Hence assertion (144) of Theorem 3 proves (121).

Now consider the case that (120a) holds. We apply Lemma 7 with $p_o := p_o$, $p := \infty$, $\overline{\delta} := n/p_o$ and initial data u_o. Hence for some $\overline{T} \in (0,T]$, the IBP has a unique solution $v \in E_{\infty,\overline{\delta},\overline{T}}$. This solution is constructed by the same Picard-Lindelöf iteration (66) which converges to the solution $u \in E_{p_1,\delta_1,T}$. Hence $u(t)=v(t)$ for all $t\in(0,\overline{T}]$ and for some $M\in(0,\infty)$

$$\|u(t)\|_{\infty} \leq m(t)^{-n/p_o} M \quad \text{for all } t\in(0,\overline{T}]. \tag{123}$$

Thus (122) proves the assertion (121) in the special case $p = p_o$ and (123) together with Lemma 13 prove the assertion (121) in the special case $p = \infty$. The general case with $p\in[p_o,\infty]$ arbitrary follows by the interpolation inequality

$$\|u\|_p \leq \|u\|_{p_o}^{p_o/p} \|u\|_{\infty}^{1-p_o/p} \quad \text{for all } p\in[p_o,\infty],\ u \in L_{\infty}(\Omega). \tag{124}$$

Thus the Lemma is proved.

Proof of Theorem 1 and 2

The following reasoning applies to Theorem 1 and 2. A local mild solution u for initial data u_o on some time interval $[0,T]$ was constructed in Lemma 13. This solution satisfies (55) and hence by Lemma 2(iv) also (56). We want to extend this solution to a global one. If one accepts the Lemma of Zorn existence of such a global extension is easy to prove. This line of reasoning is unnecessarily abstract. In this proof we will not use the Lemma of Zorn, but prefere to give a specific construction. Define the increasing sequence (T_m) and mild solution u_m for initial data u_o on the intervals $[0,T_m]$ for all $m \in N$ by the following inductive process:

Let $T_1 = T$ and let u_1 be the local mild solution for initial data u_o on the time interval $[0,T]$ given by Lemma 13.

Assume that the mild solution u_m on the interval $[0,T_m]$ is already defined. We apply Lemma 9 with initial data $u_0 := u(T_m)$. (To be quite specific, choose $u_1 = 0$ on p.41 as starting point of the iteration (66)) Hence there exists $T_{m+1} < T_m$ and a mild solution v_m for initial data $u(T_m)$ on the time interval $[T_m,T_{m+1}]$. Now u_{m+1} is defined by

$$u_{m+1}(t) = \begin{cases} u_m(t) & \text{for all } t\in(0,T_m], \\ v_m(t) & \text{for all } t\in[T_m,T_{m+1}]. \end{cases}$$

Then u_{m+1} is a mild solution for initial data u_0 on the interval $[0,T_{m+1}]$.

We take the limit $m \to \infty$. Define $T_{max} \in (0,\infty]$ by

$$\lim_{m\to\infty} T_m = T_{max} \tag{126}$$

and the mild solution u for initial data u_0 on the interval $[0,T_{max})$ by

$$u(t) = u_m(t) \quad \text{for all } t\in(0,T_m],\ m \in N. \quad \text{Since}$$

$$\sup\{\|u(t)\|_\infty \mid t\in[T_m,T_{m+1}]\} < \infty \quad \text{for all } m \in N,$$

$$\sup\{\|u(t)\|_\infty \mid t\in[T_0,T_1]\} < \infty \quad \text{for all } T_o\in(0,T_1]$$

by Lemma 9 and Lemma 13, respectively, the solution u satisfies (57). Next we prove

$$\limsup_{m\to\infty} \|u(t)\|_\infty = \infty \quad \text{if } T_{max} < \infty. \tag{127}$$

To see this, assume that (127) does not hold. Then $T_{max} < \infty$ and

$$\sup\{\|u(t)\|_\infty \mid t\in[T_1,T_{max})\} = U < \infty. \tag{128}$$

As shown in Lemma 9, the length of the interval $[T_m,T_{m+1}]$ of the m-th extension v_m depends only on q_1,q_2, $\|c\|_{q_1,q_2}$ and the function Γ from assumptions (F4)(F6) and nonincreasingly on the norm of the initial data $\|u(T_m)\|_\infty$. Hence (128) would imply

$$\sup\{T_{m+1}-T_m \mid m \in N\} = \tau > 0, \text{ which contradicts}$$

$$\lim_{m\to\infty} T_m = T_{max} < \infty.$$

Hence (127) must hold.

For the rest of the prove we distinguish Theorem 1 and 2.

To complete the proof of Theorem 1, we have to show (58). Assume that (58) is violated. Then there exists a sequence (t_k) such that

$$\lim_{k\to\infty} t_k = T_{max} < \infty \quad \text{and} \quad \sup_{k\in N} \|u(t_k)\|_{p_o} < \infty. \tag{129}$$

We choose p_1 and δ_1 as in Lemma 13 and apply Lemma 7 with $p_o := p_o$, $p := p_1$, $\delta := \delta_1$ and initial data $u_o := u(t_k)$. Hence there exists $\hat{T} \in (0,\infty)$ and there exist for all $k \in N$ mild solutions w_k for initial data $u(t_k)$ on the time intervals $[t_k, t_k+\hat{T}]$. Furtheron $w_k(\cdot, t_k+\cdot) \in E_{p_1,\delta_1,\hat{T}}$ is unique in this space. Hence for all $k \in N$

$$w_k(t) = u(t) \quad \text{for all } t \in (t_k, t_k+\hat{T}] \cap (0, T_{max}). \tag{130}$$

Take some $\hat{k}$ such that $t_{\hat{k}} + \hat{T} > T_{max}$ and define

$$\hat{u}(t) = \begin{cases} u(t) & \text{for all } t \in (0, T_{max}), \\ w_{\hat{k}}(t) & \text{for all } t \in (t_{\hat{k}}, t_{\hat{k}}+\hat{T}]. \end{cases}$$

Then $\hat{u}$ is a mild solution for initial data u_o on the interval $[0, t_{\hat{k}} + \hat{T}]$ extending the solution u. By (58) and Lemma 13 the extension $\hat{u}$ satisfies

$$\sup \{\|\hat{u}(t)\|_\infty \mid t \in [T_o, T_1]\} < \infty \quad \text{for all } T_o, T_1 \in (0, t_{\hat{k}} + \hat{T}]$$

Since $t_{\hat{k}} + \hat{T} > T_{max}$ this implies

$$\sup \{\|u(t)\|_\infty \mid t \in [T_o, T_{max})\} < \infty \quad \text{for all } T_o \in (0, T_{max}), \tag{131}$$

which contradicts (127). Hence (58) must hold.
Thus Theorem 1 is proved.

To complete the proof of Theorem 2, we show (62). Assume that (62) is violated. Then $T_{max} < \infty$ and the set

$$k(\Pi) = \{u/\|u\|_{p_o} \in L_{p_o} \mid u \in \Pi, \ u \neq 0\}$$

defined by (61) is precompact in the space L_{p_o}. Nevertheless there exists a sequence (t_k) such that (129) holds. As in Lemma 13 we choose p and δ ($p=p_1$, $\delta=\delta_1$) such that assumption (77) of Lemma 8 holds. We apply Lemma 8 to initial data $u_o := u(t_k)$ for all $k \in N$. Define

$$\ddot{\Pi} = \{u(t_k) \in L_{p_o} \mid k \in N\}.$$

Then $\ddot{\Pi} \subset L_{p_o}$ is bounded in L_{p_o} and

$$k(\ddot{\Pi}) = \{u(t_k)/\|u(t_k)\|_{p_o} \mid k \in N,\ u(t_k) \neq 0\}$$

is precompact in L_{p_o}. Hence by Lemma 8 there exists $\hat{T} \in (0,\infty)$ and there exist for all $k \in N$ mild solutions w_k for initial data $u(t_k)$ on the time intervals $[t_k, t_k+\hat{T}]$. As in the proof of Theorem 1,(130) holds and the extension $\hat{u}$ is constructed. We arrive at (131) which contradicts (127). Hence (62) must hold. Thus Theorem 2 is proved.

The following Corollary summarizes the precise results concerning uniqueness and maximality of T_{max} that one can derive from the Lemmas above. We restrict ourselves to the case of Theorem 1, because the weaker results in the case of Theorem 2 are complicated to state. Under the assumptions of Theorem 1 we state some definitions:

Let the number $p^* \in [0,\infty)$ be given by

$$\begin{aligned} &p^* = 0 && \text{if } \gamma = 1, \\ &n/q_1 + 1/q_2 + (\gamma-1)n/p^* = 1 && \text{if } \gamma > 1. \end{aligned} \tag{132}$$

Define the set ℓ by

$$\ell = \{(p,\delta) \mid p \in [1,\infty],\ \delta \in [0,\infty) \text{ such that (a)(b)(c) hold}\}, \text{ where}$$

(a) $1/q_1 + \gamma/p \leq 1$, (b) $1/q_2 + \gamma\delta < 1$, (c) $n/p_o - n/p = \delta$.

We need the linear function spaces (in which no topology is introduced)

$$X_{p_o,T} = \{u: (x,t) \in \Omega \times (0,T] \to u(x,t) \in R \mid u \in E_{p,\delta,T} \text{ for some } (p,\delta) \in \ell\},$$

$$Y_{p_o,T} = \{u: (x,t) \in \Omega \times (0,T] \to u(x,t) \in R \mid u \in E_{p,\delta,T} \text{ for all } (p,\delta) \in \ell\}.$$

<u>Corollary of Theorem 1</u> (Uniqueness and maximality)

Make the assumptions of Theorem 1 and let u be the global mild solution of the initial-boundary value problem (1)(1a)(1b) given there. Then

$$\lim_{t \to T_{max}} \|u(t)\|_p = \infty \quad \text{for all } p \in [1,\infty] \cap (p^*,\infty] \text{ if } T_{max} < \infty. \tag{133}$$

If $p^* > 1$, $T_{max} < \infty$ and the set

$$k(\Pi) = \{u(t)/\|u(t)\|_{p^*} \mid t \in (0,T_{max})\}$$

is precompact in the space $L_{p^*}(\Omega)$, then

$$\lim_{t \to T_{max}} \|u(t)\|_{p^*} = \infty. \tag{134}$$

Let $v \in X_{p_0,T}$ be any solution of the integral equation

$$v(t) = S(t)u_0 + \int_0^t S(t-s)F(s,v(s))\, ds \quad \text{for all } t \in (0,T]. \tag{135}$$

Then (v1)(v2)(v3) hold.

(v1) v is a L_{p_0}-mild solution (hence a mild solution) of the IBP.

(v2) $v \in Y_{p_0,T}$.

(v3) $T < T_{max}$ and v is the restriction to the interval $(0,T]$ of the global mild solution u given in Theorem 1.

<u>Proof</u>

To show (133) choose $t_0 \in (0,T_{max})$ arbitrary and apply Theorem 1 with $p_0 := p$ occuring in (133) and initial data $u_0 := u(t_0)$. To show (134) we apply Theorem 2 with $p_0 := p^*$ and again $u_0 := u(t_0)$.

Let $v \in X_{p_0,T}$ satisfy (135). By definition of the space $X_{p_0,T}$ there exists $(p,\delta) \in \ell$ such that the assumption (63) of Lemma 7 holds. Hence estimates (73) and (76) show that (c) and (a) in the Definition 2 p.49 of a L_{p_0}-mild solution hold. Thus the integral equation (135) implies that v is a L_{p_0}-mild solution in the sense of Definition 2. Thus (v1) is shown. Next we prove (v3).

Let u be the global mild solution of the IBP given in Theorem 1. Define the "splitting time" $\ddot{T} \in [0,T_{max}) \cap [0,T]$ by

$$\ddot{T} = \max\{t \mid u(\tau) = v(\tau) \quad \text{for all } \tau \in (0,t]\}. \tag{136}$$

If $t \in (0,T_{max}) \cap (0,T]$ is so small that

$$K_{31} m(t)^{\varepsilon} \|c\|_{q_1,q_2} \|1+|u|+|v|\|^{\gamma-1}_{p,\delta,t} \leq 1/2,$$

then estimate (75) implies $\|u-v\|_{p,\delta,t} = 0$. Hence $\ddot{T} \geq t > 0$.

Next we assume that $0 < \ddot{T} < \min\{T_{max}, T\}$ and derive a contradiction. Lemma 7 is applied with $p_o := p$, $p := p$, $\delta := 0$ and initial data $u_o := u(\ddot{T})$. By estimate (75) there exists some $s \in (0,\infty)$ such that

$$\|(u-v)(.,\ddot{T}+.)\|_{p,0,s} \leq (1/2)\ \|(u-v)(.,\ddot{T}+.)\|_{p,0,s}$$

and hence $u(\tau) = v(\tau)$ for all $\tau \in (0,\ddot{T}+s]$ contradicting the definition (136) of the splitting time $\ddot{T}$. Hence we get $\ddot{T} = \min\{T_{max}, T\}$ and

$$u(t) = v(t) \quad \text{for all } t \in (0,T_{max}) \cap (0,T]. \tag{137}$$

If $T \geq T_{max}$, (137) would imply $u \in E_{p,\delta,T}$ contradicting (133). Hence $T < T_{max}$ and the proof of (v3) is completed.

It remains to show (v2). For all $(p,\delta) \in \ell$ assumption (63) of Lemma 7 is satisfied. Hence for some $T_o \in (0,\infty)$ the Picard-Lindelöf iteration (66) yields a local solution $w \in E_{p,\delta,T}$ of the IBP. Since the iteration (66) is independent of $(p,\delta) \in \ell$, we get $w(t) = u(t)$ for all $t \in (0,T_o)$. Hence (v2) follows from (v3) and (57). Thus the Corollary is proved.

Remark

The Corollary states precisely what the word "maximal" in Theorem 1 means. Indeed, we have shown that the solution u from Theorem 1 cannot be extended within the space $X_{p_o,\infty}$. On the other hand, it may be that the solution u can be extended as a mild or L_r-mild solution in the sense of Definition 1,p.29 or 2,p.49. If the reader accepts the Lemma of Zorn, there exists a maximal extension in this sense, too. But this extension cannot be constructed by the tools from above. In view of the nonexistence and nonuniqueness results of Weissler [22,67] and the remarks in the introduction p.10, one might guess that really pathological cases can occur.

Definition 3

Let $p_o \in [1,\infty]$. A set $D \subset L_{p_o}(\Omega)$ is called generating if and only if

(a) $D \subset L_\infty(\Omega)$

(b) For all $u_o \in L_{p_o}(\Omega)$ there exists a sequence (u_{om}) in D such that

$$\|u_{om}\|_{p_o} \leq \|u_o\|_{p_o} \quad \text{for all } m \in N \quad \text{and} \quad \lim_{m\to\infty} \|u_{om} - u_o\|_1 = 0. \tag{138}$$

For example the set of regular initial data p.20 or the set $D = C_o^\infty(\overline{\Omega})$ are generating.

<u>Theorem 3</u> (Existence results exploiting a priori estimates)

Let $q_1, q_2 \in [1,\infty]$, $r_1, r_2 \in (0,\infty]$, $\gamma \in [1,\infty)$ satisfy

$$n/q_1 + 1/q_2 + (\gamma-1)(n/r_1 + 1/r_2) < 1 , \tag{139}$$

$$1/q_1 + (\gamma-1)/r_1 \leq 1 . \tag{140}$$

Let $p_o \in [1,\infty]$ be arbitrary.

Let F satisfy (FO), the two-sided global bound (F1) and the global Lip. condition (F3) with q_1, q_2, γ as stated above.

Assume that there exist $T \in (0,\infty)$, a generating set D and an increasing function $G\colon u \in [0,\infty) \to G(u) \in [0,\infty)$ such that every mild solution u of the IBP for initial data $u_o \in D$ on any time interval $[0,\tilde{T}] \subset [0,T]$ (where $\tilde{T}$ may depend on u_o and u) satisfies an a priori estimate

$$|||u|||_{r_1, r_2, \tilde{T}} \leq G(\|u_o\|_{p_o}) . \tag{141}$$

Then for all initial data $u_o \in L_{p_o}(\Omega)$ the initial-boundary value problem problem (1)(1a)(1b) has a mild solution u on the time interval $[0,T]$ (which was already specified above!). This solution u satisfies (141). For $p \in [1,\infty]$, $\delta \in [0,\infty)$ chosen according to Lemma 15 we have $u \in E_{p,\delta,T}$ and the solution u is unique in this space. Furtheron

$$\lim_{t\to 0} \|u(t) - S(t)u_o\|_{p_o} = 0, \tag{142}$$

$$\lim_{t\to 0} \|u(t) - u_o\|_{p_o} = 0 \quad \text{if assumption (A) p.20 holds.} \tag{143}$$

The solution u can be extended on a maximal time interval $[0,T_{max}) \supset [0,T]$ such that

$$\sup \{m(t)^\delta \|u(t)\|_p \mid t \in (0,T_2]\} < \infty \quad \text{for all } T_2 \in (0,T_{max}), \; p \in [p_o,\infty],\ \delta = n/p_o - n/p. \tag{144}$$

$$\lim_{t\to T_{max}} \|u(t)\|_p = \infty \quad \text{for all } p \in [1,\infty] \cap (p^*,\infty] \text{ if } T_{max} < \infty, \tag{145}$$

where p^* is defined by (132).

$$\int_{T_{max}-\tau}^{T_{max}} \|u(t)\|_{r_1}^{r_2} \, dt = \infty \quad \text{for all } \tau > 0, \text{ if } r_2 < \infty. \tag{146}$$

If the initial data satisfy $u_o \geqq 0$ and the nonlinearity satisfies the positivity assumption (F7), the solution satisfies $u \geqq 0$.

Remark

The strong Lipschitz condition (F3) is needed to establish continuous dependence of the solution u on the initial data u_o. To show (145) or (146) the Lipschitz condition is not needed. This result will be given more precisely in Theorem 5 p.77.

The proof of Theorem 3 uses some Lemmas:

Lemma 15 establishes the Lipschitz continuous dependence of the solution on the initial data and shows in which sense the initial data are approximated for $t \to 0$. Here the primary a priori estimate (141) is used.

Lemma 16 constructs the chains of exponents $p_o < p_1 < \ldots < p_k = \infty$ and (δ_i) $i = 1 \ldots k$ needed for the successive application of Lemma 15 and lateron Lemma 19, too.

Lemma 17 estimates $\|u(t)\|_\infty$ by successive application of Lemma 15.

Lemma 15

Let $p_o, q_1, q_2 \in [1,\infty]$, $p \in [p_o,\infty]$, $r_1, r_2 \in (0,\infty]$, $\gamma \in [1,\infty)$, $\delta \in [0,\infty)$, $\varepsilon \in (0,\infty)$ satisfy (147)(148)(149)(150).

$$n/q_1 + 1/q_2 + (\gamma-1)(n/r_1 + 1/r_2) < 1 - \varepsilon \tag{147}$$

$$1/q_1 + (\gamma-1)/r_1 + 1/p \leqq 1 \tag{148}$$

$$1/q_2 + (\gamma-1)/r_2 + \delta < 1 - \varepsilon \tag{149}$$

$$n/p_o - n/p = \delta \tag{150}$$

Let F satisfy (F0), the two-sided bound (F1) and the Lipschitz condition (F3) with q_1, q_2, γ from above.

Assume that u_i for $i = 1,2$ are mild solutions of the IBP for initial data $u_{oi} \in L_{p_o}$ on some time interval $[0,T]$ which satisfy an a priori bound

$$|||u_i|||_{r_1,r_2,T} < \infty \quad \text{for } i = 1,2. \tag{151}$$

Then there exists a constant $K_{41}(q_1,q_2,r_1,r_2,p_o,p,\gamma,\varepsilon,N,A_o)$ such that the following estimates hold for $i = 1,2$:

$$\|u_i\|_{p,\delta,T} \leq K_{41}\|u_{oi}\|_{p_o} + K_{41}m(T)^{\varepsilon}\|c\|_{q_1,q_2}|||1+|u_i|\,|||^{\gamma-1}_{r_1,r_2,T}\|1+|u_i|\,\|_{p,\delta,T} \tag{152}$$

$$\|u_1-u_2\|_{p,\delta,T} \leq K_{41}\|u_{o1}-u_{o2}\|_{p_o} + \tag{153}$$

$$+ K_{41}m(T)^{\varepsilon}\|c\|_{q_1,q_2}|||1+|u_1|+|u_2|\,|||^{\gamma-1}_{r_1,r_2,T}\|u_1-u_2\|_{p,\delta,T}$$

$$\|u_i(t)-S(t)u_{oi}\|_{p_o} \leq K_{41}m(t)^{\varepsilon}\|c\|_{q_1,q_2}|||1+|u_i|\,|||^{\gamma-1}_{r_1,r_2,t}\|1+|u_i|\,\|_{p,\delta,t} \tag{154}$$

$$\int_0^t\|F(s,u_i(s))\|_1 ds \leq K_{41}(t+t^{\varepsilon})\,|c|_{q_1,q_2}\|1+|u_i|\,|||^{\gamma-1}_{r_1,r_2,t}\|1+|u_i|\,\|_{p,\delta,t} \tag{155}$$

<u>Proof</u>

Choose $s_1,s_2\in[1,\infty]$ such that

$$1/q_1 + (\gamma-1)/r_1 + 1/p \leq 1/s_1, \quad 1/q_2 + (\gamma-1)/r_2 \leq 1/s_2, \qquad (156),(157)$$

$$n/s_1 + 1/s_2 - n/p + \varepsilon < 1 \quad \text{and} \quad 1/s_2 + \delta + \varepsilon \leq 1. \tag{158}$$

Subtracting the two integral equations

$$u_i(t) = S(t)u_{oi} + \int_0^t S(t-s)F(s,u_i(s))\,ds$$

for $i = 1,2$ and using the Lipschitz condition (F3),(156) and Hölder's inequality (H3)p.35 we arrive at the estimate

$$\|(u_1-u_2)(t)\|_p \leq \|S(t)\|_{p,p_o}\|u_{o1}-u_{o2}\|_{p_o} +$$

$$+ \int_0^t \|S(t-s)\|_{p,s_1}\|c(.,s)(1+|u_1|+|u_2|)^{\gamma-1}|u_1-u_2|(s)\|_{s_1}\,ds$$

$$\leq \|S(t)\|_{p,p_o}\|u_{o1}-u_{o2}\|_{p_o} + \tag{159}$$

$$+ \left[\int_0^t\|S(t-s)\|_{p,s_1}\|c(.,s)\|_{q_1}\|(1+|u_1|+|u_2|)(s)\|^{\gamma-1}_{r_1}m(s)^{-\delta}\,ds\right]\|u_1-u_2\|_{p,\delta,T}.$$

The first summand is estimated by Lemma 3,(30) and the second by Lemma 6 with p, $s_1,s_2,\delta,\varepsilon$ from above, $\alpha := \delta$ and

$$f(s) := \|c(.,s)\|_{q_1}\|(1+|u_1|+|u_2|)(s)\|^{\gamma-1}_{r_1}.$$

By (158) assumption (51) of Lemma 6 is satisfied. Hence (159) implies

$$\|u_1-u_2\|_{p,\delta,T} \leq K_{42}\|u_{o1}-u_{o2}\|_{p_o} + K_{42}m(T)^{\varepsilon}p_{s_2}(f)\|u_1-u_2\|_{p,\delta,T}. \qquad (160)$$

To estimate $p_{s_2}(f)$ (see Lemma 6 for the definition) we use (157) and Hölder's inequality. Hence we get

$$p_{s_2}(f) \leq \|c\|_{q_1,q_2}|\!|\!| 1+|u_1|+|u_2| |\!|\!|^{\gamma-1}_{r_1,r_2,T}. \qquad (161)$$

Now (160)(161) imply the assertion (153). Estimate (152) is shown by an analogous argument using the bound (F1) for F instead of the Lipschitz condition (F3).

The arguments to prove (154) are quite similar. From the integral equation we get by the bound (F1),(156) and Hölder's inequality

$$\|u(t) - S(t)u_o\|_{p_o} \leq \qquad (162)$$

$$\left[\int_0^t \|S(t-s)\|_{p_o,s_1}\|c(.,s)\|_{q_1}\|1+|u|(s)\|^{\gamma-1}_{r_1}m(s)^{-\delta}\,ds\right]\|1+|u|\|_{p,\delta,t}.$$

We apply Lemma 6 with s_1,s_2 from above, $\alpha := \delta$, $\delta := 0$, $\varepsilon := \varepsilon$ and

$$\tilde{f}(s) := \|c(.,s)\|_{q_1}\|1+|u|(s)\|^{\gamma-1}_{r_1}.$$

We estimate $p_{s_2}(\tilde{f})$ by (152) and Hölder's inequality:

$$p_{s_2}(\tilde{f}) \leq \|c\|_{q_1,q_2}|\!|\!|1+|u||\!|\!|^{\gamma-1}_{r_1,r_2,T}. \qquad (163)$$

Now (162)(163) imply the assertion (154). By a modification of (162)

$$\int_0^t\|F(s,u(s)\|_1 ds \leq \int_0^t \tilde{f}(s)m(s)^{-\delta}ds\|1+|u|\|_{p,\delta,t} \leq K(t+t^{\varepsilon})p_{s_2}(\tilde{f})\|1+|u|\|_{p,\delta,t}.$$

Hence (163) implies (155). Thus the Lemma is proved.

Lemma 16

Let $p_o\in[1,\infty]$ be arbitrary and $q_1,q_2\in[1,\infty]$, $r_1,r_2\in(0,\infty]$, $\gamma\in[1,\infty)$, $\beta\in[0,1]$

$$n/q_1 + 1/q_2 + \gamma(1-\beta)(n/r_1 + 1/r_2) < 1, \qquad (164)$$

$$1/q_1 + \gamma(1-\beta)/r_1 \leq 1, \qquad (165)$$

$$\gamma\beta \leq 1. \qquad (166)$$

Then there exist $k \in N$, $\varepsilon \in (0,1)$ and finite sequences (p_i) and (δ_i) such that

$$p_o < p_1 < p_2 < \dots < p_k = \infty,$$

$$\delta_i \in (0,1) \quad \text{for } i = 1 \dots k$$

and (167)(168)(169) hold.

$$1/q_1 + \gamma(1-\beta)/r_1 + \gamma\beta/p_i \leq 1 \qquad \text{for all } i = 1\dots k \tag{167}$$

$$1/q_2 + \gamma(1-\beta)/r_2 + \gamma\beta\delta_i < 1 - \varepsilon \text{ for all } i = 1\dots k \tag{168}$$

$$n/p_{i-1} - n/p_i = \delta \qquad \text{for all } i = 1\dots k \tag{169}$$

<u>Proof</u>

Let $\beta \in (0,1]$ for otherwise the result is trivial. The important point is to choose $p_1 \in (p_o, \infty]$ such that (167)(168) hold with δ_1 given by (169). These conditions can be written more systematically

$$\begin{aligned} 0 \leq \gamma\beta n/p_1 &< \gamma\beta n/p_o, \\ \gamma\beta n/p_1 &\leq n - n/q_1 - \gamma(1-\beta)/r_1, \end{aligned} \tag{170}$$

$$1/q_2 + \gamma(1-\beta)/r_2 + \gamma\beta n/p_o - 1 + \varepsilon < \gamma\beta n/p_1.$$

There exists $p_1 \in [1,\infty]$ satisfying (170) if and only if

$$\begin{aligned} n/q_1 + 1/q_2 + \gamma(1-\beta)(n/r_1 + 1/r_2) &< 1 + n(1 - \gamma\beta/p_o) - \varepsilon, \\ 1/q_1 + \gamma(1-\beta)/r_1 &\leq 1, \\ 1/q_2 + \gamma(1-\beta)/r_2 &< 1 - \varepsilon. \end{aligned} \tag{171}$$

Under the assumptions of the Lemma (171) holds. Hence there exist p_1 and δ_1 satisfying (167)(168)(169) for $i = 1$. Now choose $p_2 < p_3 < \dots < p_k = \infty$ and (δ_i) for $i = 2\dots k$ such that

$$\delta_i = n/p_{i-1} - n/p_i \leq \delta_1 \quad \text{for } i = 2,\dots k.$$

Then all assertions of the Lemma are satisfied.

Lemma 17

Let $p_o, q_1, q_2 \in [1,\infty]$, $r_1, r_2 \in (0,\infty]$, $\gamma \in [1,\infty)$ and the function F satisfy the same hypotheses as in Theorem 3. Let $u_o \in L_\infty(\Omega)$ and assume that the mild solution of the IBP for initial data u_o given by Theorem 1 restricted to the time interval $[0,T]$ (i.e. $T < T_{max}$) satisfies the estimate

$$\|c\|_{q_1,q_2} |||1+|u|\,|||^{\gamma-1}_{r_1,r_2,T} \leq C. \tag{172}$$

Then there exist ε, K_{43}, $T_o \in (0,\infty)$ depending on the quantities in the brackets:

$$\varepsilon = \varepsilon(p_o,q_1,q_2,r_1,r_2,\gamma,N), \quad K_{43} = K_{43}(p_o,q_1,q_2,r_1,r_2,\gamma,N,A_o),$$

$$T_o = T_o(p_o,q_1,q_2,r_1,r_2,\gamma,N,A_o,C) \quad \text{such that}$$

$$\|u(t)\|_\infty \leq K_{43} m(t)^{-n/p_o} (\|u_o\|_{p_o} + m(t)^\varepsilon C) \text{ for all } t \in (0,T_o] \cap (0,T] \tag{173}$$

Proof

By (139) there exists $\varepsilon \in (0,1)$ such that (147) holds. First we consider the simpler special case $p_o = \infty$. We apply Lemma 15 with $\gamma, q_1, q_2, r_1, r_2$ from above, $p := \infty$, $\delta := 0$. Hence (152) yields

$$\|u\|_{\infty,0,t} \leq K_{44}\|u_o\|_\infty + K_{44} m(t)^\varepsilon C \|u\|_{\infty,0,t} \quad \text{for all } t \in (0,T]. \tag{174}$$

Choose $T_o \in (0,T]$ small enough such that

$$K_{44} m(T_o)^\varepsilon C \leq 1/2. \tag{175}$$

Since $|u|_{\infty,0,T_o} < \infty$ by (57) Theorem 1, estimate (174) and (175) imply

$$\|u\|_{\infty,0,t} \leq 2K_{44}\|u_o\|_\infty \quad \text{for all } t \in (0,T_o],$$

which proves the assertion (173) for the special case $p_o = \infty$.

Now assume $p_o \in [1,\infty)$. Define $\beta := 1/\gamma$ and let $p_o < p_1 < \ldots < p_k = \infty$ and (δ_i) $i = 1\ldots k$ be the sequences constructed by Lemma 16. Choose $\varepsilon \in (0,1)$ small enough such that (147) and (168) hold. Sequentially for $i = 1\ldots k$ we apply Lemma 15 with $p_o := p_{i-1}$, $p := p_i$, $\delta := \delta_i$, $q_1, q_2, r_1, r_2, \gamma, \varepsilon$ from above and initial data $u_o := u(t)$ for $t \in [0,T)$. Hence (152) yields

$$\|u(.,t+.)\|_{p_i,\delta_i,h} \leq K_i \|u(t)\|_{p_{i-1}} + K_i m(h)^\varepsilon C (1 + \|u(.,t+.)\|_{p_i,\delta_i,h}) \tag{176}$$
$$\text{for all } i = 1\ldots k,\ t \in [0,T),\ h \in (0,T-t].$$

We choose $T_o > 0$ small enough such that

$$\max_{1\le i\le k} K_i \, m(T_o)^{\varepsilon} C \le 1/2. \tag{177}$$

By (57) Theorem 1, both sides of the inequality (176) are finite. Hence (176)(177) with $t = ih-h$ imply

$$\begin{aligned} m(h)^{\delta_i} \| u(ih) \|_{p_i} &\le \| u(.,ih-h+.) \|_{p_i,\delta_i,h} \qquad \text{for all } i = 1...k, \\ &\le 2K_i [\| u(ih-h) \|_{p_{i-1}} + m(h)^{\varepsilon} C] \qquad h\in(0,T_o]\cap(0,T/i). \end{aligned} \tag{178}$$

By induction on i this implies

$$m(h)^{(\delta_1+...+\delta_i)} \| u(ih) \|_{p_i} \le 2^i K_1 ... K_i [\| u_o \|_{p_o} + i m(h)^{\varepsilon} C] \tag{179}$$
$$\text{for all } i = 1...k, \; h\in(0,T_o]\cap(0,T/i].$$

By (169) we get $\delta_1+\delta_2+...+\delta_k = n/p_o$. Hence for $i = k$ (179) yields the assertion (173). Thus the Lemma is proved.

<u>Proof of Theorem 3</u>

For initial data $u_o \in L_{p_o}$ there exists a sequence (u_{om}) in the set $D \subset L_\infty$ such that according to Definition 3 p.56

$$\lim_{m\to\infty} \| u_{om} - u_o \|_1 = 0 \quad \text{and} \quad \| u_{om} \|_{p_o} \le \| u_o \|_{p_o} \quad \text{for all } m \in N. \tag{180}$$

Theorem 1 proves the existence of mild solutions u_m of the IBP for initial data u_{om} on the time intervals $[0,T_m)$. Here $T_m = T_{max,m}$ is the maximal existence time from Theorem 1. First we show

$$\inf \{ T_m | \; m \in N \} > 0. \tag{181}$$

By assumption (141) we have the a priori estimates

$$||| u_m |||_{r_1,r_2,\tilde{T}} \le G(\| u_o \|_{p_o}) \quad \text{for all } \tilde{T}\in(0,T]\cap(0,T_m). \tag{182}$$

Hence we can apply Lemma 17 with

$$C = \| c \|_{q_1,q_2} [1 + 2G(\| u_o \|_{p_o})]^{\gamma-1}$$

and get the estimate

$$\|u_m(t)\|_\infty \le K_{45} m(t)^{-n/p_0}[\|u_0\|_{p_0} + m(t)^{\varepsilon} C] \quad \text{for all } m \in N, \; t \in (0,T_0] \cap (0,T] \cap (0,T_m). \tag{183}$$

Here $T_0 \in (0,T]$ is given in Lemma 17 by condition (177), hence T_0 depends only on the exponents, the operator A_0 and on C given above, but not on $m \in N$ or T. Since by (58) from Theorem 1

$$\lim_{t\to T_{max}} \|u_m(t)\|_\infty = \infty \quad \text{if } T_m < \infty,$$

estimate (183) implies $T_m > \min\{T,T_0\}$ for all $m \in N$ and hence (181).

Next we prove convergence of the sequence u_m in a suitable Banach space. We apply Lemma 16 with $\beta := 1/\gamma$ and $\tilde{p}_0 := 1$. Hence there exist $\tilde{p} = p_1$, $\tilde{\delta} = \delta_1$ and $\tilde{\varepsilon} \in (0,1)$ such that (147)(148)(149)(150) hold with "~". Since these are the assumptions of Lemma 15, estimate (153) yields

$$\|u_n - u_m\|_{\tilde{p},\tilde{\delta},\tilde{T}} \le K_{46}\|u_{on} - u_{om}\|_1 + \tag{184}$$

$$+ K_{46} m(T)^{\tilde{\varepsilon}} \|c\|_{q_1,q_2} \||1+|u_n|+|u_m|\||^{\gamma-1}_{r_1,r_2,T} \|u_n - u_m\|_{\tilde{p},\tilde{\delta},\tilde{T}}$$

$$\text{for all } n,m \in N,\; \tilde{T} \in (0,T] \cap (0,T_0].$$

Choose $\tilde{T} \in (0,T] \cap (0,T_0]$ small enough such that

$$K_{46} m(\tilde{T})^{\tilde{\varepsilon}} \|c\|_{q_1,q_2} (1+2G(\|u_0\|_{p_0}))^{\gamma-1} \le 1/2. \tag{185}$$

Then (141) and (184)(185) imply

$$\|u_n - u_m\|_{\tilde{p},\tilde{\delta},\tilde{T}} \le 2K_{46}\|u_{on} - u_{om}\|_1. \tag{186}$$

Hence by (180) there exists $u \in E_{\tilde{p},\tilde{\delta},\tilde{T}}$ such that

$$\lim_{m\to\infty} \|u_m - u\|_{\tilde{p},\tilde{\delta},\tilde{T}} = 0. \tag{187}$$

From (141) and (183) we derive estimates for u. Let $t \in (0,\tilde{T}]$ be fixed. By the Theorem of Riesz-Fischer (Segal and Kunze [61] p.97) there exists a subsequence of $u_m(t)$ converging to $u(t)$ for almost all $x \in \Omega$. Hence (183) implies

$$\|u(t)\|_\infty \le K_{45} m(t)^{-n/p_0}[\|u_0\|_{p_0} + m(t)^{\varepsilon} C] \quad \text{for all } t \in (0,\tilde{T}]. \tag{188}$$

By Fatou's Lemma (Segal and Kunze [61] p.69) (182) and (187) imply

$$\|u\|_{r_1,r_2,\tilde{T}} \leq G(\|u_o\|_{p_o}). \tag{189}$$

Next we show that u is a local mild solution of the IBP for initial data u_o on the time interval $[0,\tilde{T}]$. We have to take the limit $m \to \infty$ in the integral equations

$$u_m(t) = S(t)u_{om} + \int_0^t S(t-s)F(s,u_m(s))\,ds \quad \text{for all } t\in(0,\tilde{T}]. \tag{190}$$

We apply Lemma 15 with $\tilde{p}_o := 1$ and $\tilde{p},\tilde{\delta},\tilde{\varepsilon}$. A slight modification of (155), furtheron (189) and the definition of C p.63 yield

$$\int_0^t \|F(s,u_m(s))-F(s,u(s))\|_1 ds$$

$$\leq K_{47}(t+t^{\tilde{\varepsilon}})\|c\|_{q_1,q_2}\|1+|u|+|u_m|\|^{\gamma-1}_{r_1,r_2,\tilde{T}}\|u_m-u\|_{\tilde{p},\tilde{\delta},\tilde{T}}$$

$$\leq K_{47}(t+t^{\tilde{\varepsilon}})C\|u_m-u\|_{\tilde{p},\tilde{\delta},\tilde{T}} \quad \text{for all } t\in(0,\tilde{T}].$$

Using (180) and (187) we can take $m \to \infty$ in the integral equation (190) Hence the function u satisfies the integral equation (45). Thus u is a local mild solution of the IBP in the sense of Definition 1 p.29.

Now we show that the initial data u_o are assumed in the sense (142)(143). Let $p_o \in [1,\infty]$be again the mumber occuring in the a priori estimate (141). For $p_o = \infty$, let $p_1 = \infty$, $\delta_1 = 0$. For $p_o\in[1,\infty)$ choose p_1 and δ_1 as in the proof of Lemma 17 p.62. Since T_o was chosen according to (177), estimate (178) for i = 1 implies

$$\|u_m\|_{p,\delta,h} \leq 2K_1[\|u_o\|_{p_o} + m(h)^{\varepsilon}C] \quad \text{for all } m \in N,\ h\in(0,T]\cap(0,T_o].$$

Since $\tilde{T} = \min\{T,T_o\}$ we get by using again the Theorem of Riesz-Fischer

$$\|u\|_{p,\delta,\tilde{T}} \leq 2K_1[\|u_o\|_{p_o} + m(h)^{\varepsilon}C]. \tag{191}$$

Since assumptions (147)(148)(149)(150) of Lemma 15 are satisfied for $p = p_1$, $\delta = \delta_1$, estimate (154) implies

$$\|u(t)-S(t)u_o\|_{p_o} \leq K_{41}m(t)^{\varepsilon}C(1+\|u\|_{p,\delta,t}) \quad \text{for all } t\in(0,\tilde{T}]. \tag{192}$$

Now (191)(192) imply the assertion (142). Lemma 2(iv) implies (143).

By Theorem 1 and the Corollary the local solution u can be extended to a maximal time interval $[0,T_{max})$ such that (57) and (133) = (145) hold.

To prove (144) we first consider the special cases (i) and (ii).
(i) Let $p = p_o$, $\delta = 0$. (144) follows from (57) and (142).
(ii) Let $p = \infty$, $\delta = n/p_o$. (144) follows from (57) and (188).
Now the interpolation inequality

$$\|u\|_p \leq \|u\|_{p_o}^{p_o/p} \|u\|_\infty^{(1-p_o/p)} \quad \text{for all } u \in L_\infty(\Omega), \quad p\in[p_o,\infty]$$

yields (144) for all $p\in[p_o,\infty]$, $\delta = n/p_o - n/p$.

It remains to prove that the solution u can be extended at least to the interval $[0,T]$ for which the a priori estimate (141) is available. In the special case $r_2 = \infty$ this follows from (145). Consider the case $r_2 < \infty$ and assume $T_{max} \leq T < \infty$ to derive a contradiction. The primary a priori estimate (141) implies

$$\|u\|_{r_1,r_2,T_{max}} \leq G(\|u_o\|_{p_o}) < \infty.$$

We apply Lemma 17 with C from p.63 and initial data $u(t) \in L_\infty$ for arbitrary $t \in (0,T_{max})$. Hence there exists $T_o > 0$ depending on the exponents, the operator A_o and on C, but not on $t \in (0,T_{max})$ such that by (173)

$$\|u(t+s)\|_\infty \leq K_{43} m(s)^{-n/p_o}(\|u(t)\|_{p_o} + C) \quad \text{for all } t\in(0,T_{max}), \; s \in (0,T_o]\cap(0,T_{max}-t).$$

Choosing $t = T_{max} - T_o/2$, $s \in (0,T_{max}/2)$ this estimate would imply

$$\limsup_{t \to T_{max}} \|u(t)\|_\infty < \infty$$

contradicting (145). Hence $T_{max} > T$. This argument proves (146), too.

Take initial data $u_o \geq 0$ and let F satisfy the positivity assumption (F7). One can choose the approximating sequence in (180) such that $u_{om} \geq 0$ for all $m \in N$. Then Lemma 10 proves $u_m \geq 0$ for all $m \in N$. Hence (187) implies $u \geq 0$. By Lemma 10 $u \geq 0$ remains valid after the extension of the solution u to the maximal time interval.

Thus the Theorem is proved.

Theorem 4 (Global existence and global a priori estimates)

Let $q_1, q_2 \in [1,\infty]$, $r_1, r_2 \in (0,\infty]$, $\gamma \in [1,\infty)$ satisfy

$$n/q_1 + 1/q_2 + (\gamma-1)(n/r_1 + 1/r_2) < 1, \tag{193}$$

$$1/q_1 + (\gamma-1)/r_1 < 1 \quad \text{if} \quad r_1 < \infty. \tag{194}$$

Let $p_0 \in [1,\infty]$ be arbitrary.

Let the nonlinearity F satisfy (FO), the two-sided global bound (F1) and the global Lipschitz condition (F3) with q_1, q_2, γ from above.

Assume that there exists a generating set D (see Definition 3 p.56) and an increasing function $G: u \in [0,\infty) \to G(u) \in [0,\infty)$ such that every mild solution u of the initial-boundary value problem (1)(1a)(1b) for initial data $u_0 \in D$ on any time interval $[0,T]$ satisfies an a priori estimate

$$|||u|||_{r_1, r_2, T} \leq G(\|u_0\|_{p_0}). \tag{197}$$

(note that we assume this estimate for every $T > 0$ for which it makes sense, not only for some T given from the beginning as in Theorem 3)

Then for every initial data $u_0 \in L_{p_0}(\Omega)$ the initial-boundary value problem (1)(1a)(1b) has a global mild solution u on the time interval $[0,\infty)$. Furthermore this solution satisfies

$$\lim_{t\to 0} \|u(t) - S(t)u_0\|_{p_0} = 0 \tag{198}$$

$$\lim_{t\to 0} \|u(t) - u_0\|_{p_0} = 0 \quad \text{if assumption (A) p.20 holds} \tag{199}$$

$$\sup_{0<t<\infty} m(t)^{\delta} \|u(t)\|_p < \infty \quad \text{for all } p \in [p_0,\infty],\ \delta = n/p_0 - n/p \tag{200}$$

There exist exponents $\alpha \in (0,\infty)$, $\varepsilon \in (0,1)$ depending only on the exponents $p_0, q_1, q_2, r_1, r_2, \gamma, N$ and a constant K_{48} depending on these exponents and on the operator A_0 such that

$$\|u(t)\|_\infty \leq K_{48} m(t)^{-n/p_0}[\|u_0\|_{p_0} + m(t)^{\varepsilon}(\Phi + \Phi^{\alpha})], \text{ where} \tag{201}$$

$$\Phi = (1 + \|c\|_{q_1,q_2})(1 + G(\|u_0\|_{p_0})).$$

Estimate (201) holds with $\alpha = \gamma$, if one assumes

$$n/q_1 + 1/q_2 + \gamma(n/r_1 + 1/r_2) < 1 \quad \text{and} \quad 1/q_1 + \gamma/r_1 \leq 1. \tag{202}$$

If the initial data satisfy $u_o \geqq 0$ and the nonlinearity satisfies the positivity assumption (F7), the solution given above satisfies $u \geqq 0$.

For the proof of Theorem 4 we need some Lemmas:

Lemma 18 states the technical basis of the "feedback"-argument.

Lemma 19 derives stronger a priori estimates from the primary estimate (197) by means of the "feedback"-argument. A brief sketch of the idea was given in the introduction p.6.

Lemma 20 constructs L_∞-estimates from the primary a priori estimate by successive application of Lemma 19 for the chain of exponents (p_i), (δ_i) $i = 1...k$ given by Lemma 16.

Note that Lemma 19 and 20 do not use any Lipschitz condition. This advantage will be exploited in Theorem 5 and 6. Indeed, these Lemmas are the heart of the whole monograph.

Lemma 18

Assume that $a,b,y\in[0,\infty)$, $\sigma\in(0,1)$. Then

$$y \leqq a + by^\sigma \quad \text{implies} \quad y \leqq a/(1-\sigma) + b^{1/(1-\sigma)}. \tag{203}$$

Proof

For all $a\in(0,\infty)$ the equation

$$x = a + bx^\sigma \tag{204}$$

has exactly one solution $x = x(a) \in (0,\infty)$. Hence (204) defines a function x: $a\in(0,\infty) \to x(a)\in(0,\infty)$. Differentiating (204) with respect to a yields after some calculations

$$x'(a) = \frac{x(a)}{(1-\sigma)x(a) + \sigma a}. \tag{205}$$

Since $x(a) > a$, this implies

$$1 < x'(a) < 1/(1-\sigma) \quad \text{for all } a\in(0,\infty). \tag{206}$$

$B \equiv b^{1/(1-\sigma)} < x(a)$ for all $a\in(0,\infty)$ and $\lim_{\alpha\to 0} x(\alpha)$ exists by (206), thus (204) implies $\lim_{\alpha\to 0} x(\alpha) = B$. Now (206) implies for all $a\in(0,\infty)$

$$x(a) \leqq \lim_{\alpha\to 0} x(\alpha) + \max_{\alpha\geqq 0} x'(\alpha)\, a = b^{1/(1-\sigma)} + a/(1-\sigma).$$

Since $y \leqq x(a)$ this proves the assertion (203).

<u>Lemma 19</u>

Let $p_o,q_1,q_2 \in [1,\infty]$, $p \in [p_o,\infty]$, $r_1,r_2 \in (0,\infty]$, $\beta \in [0,1]$, $\gamma \in [1,\infty)$, $\delta \in [0,\infty)$, $\varepsilon \in (0,1)$ satisfy

$$n/q_1 + 1/q_2 + \gamma(1-\beta)(n/r_1 + 1/r_2) < 1 + (1-\gamma\beta)n/p - \varepsilon \tag{207}$$

$$1/q_1 + \gamma(1-\beta)/r_1 + \gamma\beta/p \leq 1 \tag{208}$$

$$1/q_2 + \gamma(1-\beta)/r_2 + \gamma\beta\ \delta < 1 - \varepsilon \tag{209}$$

$$\gamma\beta < 1 \tag{210}$$

$$n/p_o - n/p = \delta \tag{211}$$

Let the function F satisfy (FO) and the two-sided local bound (F1) with q_1,q_2,γ from above. If additionally $u_o \geqq 0$, $u \geqq 0$, the one-sided bound (F2) instead of (F1) is sufficient. No Lipschitz condition is needed.

Let the mild solution u of the IBP for initial data $u_o \in L_{p_o}$ on the interval $[0,T]$ satisfy the a priori estimate

$$|||u|||_{r_1,r_2,T} < \infty. \tag{212}$$

Furtheron, for $\beta > 0$ we assume $\|u\|_{p,\delta,T} < \infty$. We define

$$U = \|c\|_{q_1,q_2} |||1+|u|\,|||^{\gamma(1-\beta)}_{r_1,r_2,T}. \tag{213}$$

Under the assumptions above there exists a constant K_{49} depending on the exponents $p,p_o,q_1,q_2,r_1,r_2,\beta,\delta,\varepsilon,N$ and the operator A_o (but not on u_o, U or T) such that

$$\|u\|_{p,\delta,T} \leq K_{49}\|u_o\|_{p_o} + K_{49}m(T)^{\varepsilon}(U + U^{1/(1-\gamma\beta)}). \tag{214}$$

In the case that the two-sided bound (F1) holds we get furtheron

$$\|u(t) - S(t)u_o\|_{p_o} \leq K_{49}m(t)^{\varepsilon}U\|1+|u|\,\|^{\gamma\beta}_{p,\delta,t} \quad \text{for all } t\in(0,T] \tag{215}$$

$$\text{and} \quad |||\tilde{F}|||_{1,1,T} \leq K_{49}(T+T^{\varepsilon})\,U[\|u_o\|_{p_o} + U + U^{1/(1-\gamma\beta)}] \tag{216}$$

for the nonlinearity $\tilde{F} = \tilde{F}(x,t) = F(x,t,u(x,t))$.

Proof

There exist $s_1, s_2 \in [1,\infty]$ such that

$$n/s_1 + 1/s_2 \leq 1 + n/p - \varepsilon, \tag{217}$$

$$1/s_2 + \gamma\beta\delta < 1 - \varepsilon, \tag{218}$$

$$1/q_1 + \gamma(1-\beta)/r_1 + \gamma\beta/p \leq 1/s_1, \tag{219}$$

$$1/q_2 + \gamma(1-\beta)/r_2 \leq 1/s_2. \tag{220}$$

The integral equation (45) implies the estimate

$$\|u(t)\|_p \leq \|S(t)\|_{p,p_0}\|u_0\|_{p_0} + \int_0^t \|S(t-\tau)\|_{p,s_1}\|F(\tau,u(\tau))\|_{s_1}\, d\tau. \tag{221}$$

We apply (F1) and split the nonlinear term into three factors. Then (219) and the Hölder inequality (H3) p.35 imply that for all $t\in(0,T]$

$$m(t)^{\delta}\|u(t)\|_p \leq m(t)^{\delta}\|S(t)\|_{p,p_0}\|u_0\|_{p_0} + \tag{222}$$

$$+ m(t)^{\delta}\left[\int_0^t \|S(t-\tau)\|_{p,s_1}\|c(.,\tau)\|_{q_1}\|1+|u|(\tau)\|_{r_1}^{\gamma(1-\beta)} m(\tau)^{-\gamma\beta\delta}d\tau\right]\|1+|u|\|_{p,\delta,T}^{\gamma\beta}.$$

The first summand is estimated by (30) Lemma 3(i). To estimate the se-second summand we apply Lemma 6 with $p, s_1, s_2, \delta, \varepsilon$ from above, $\alpha := \gamma\beta\delta$,

$$f(\tau) := \|c(.,\tau)\|_{q_1}\|1+|u|(\tau)\|_{r_1}^{\gamma(1-\beta)}.$$

Hence (53) and (222) imply

$$\|u\|_{p,\delta,T} \leq K_{50}\|u_0\|_{p_0} + K_{50}m(T)^{\varepsilon}p_{s_2}(f)\ \|1+|u|\|_{p,\delta,T}^{\gamma\beta}. \tag{223}$$

The functional $p_{s_2}(f)$ from Lemma 6 p.31 can be estimated using (220) and Hölder's inequality:

$$p_{s_2}(f) = \|c\|_{q_1,q_2}\|1+|u|\|_{r_1,r_2,T}^{\gamma(1-\beta)} = U \tag{224}$$

where U was already defined by (213). Now (223)(224) imply

$$\|u\|_{p,\delta,T} \leq K_{50}\|u_0\|_{p_0} + K_{50}m(T)^{\varepsilon}U\|1+|u|\|_{p,\delta,T}^{\gamma\beta}. \tag{225}$$

In the special case $\beta = 0$ this is already the assertion (214).
For $\beta\in(0,1]$ we apply the "feedback"-argument to (225). Let a,b,y and σ in Lemma 18 be given by

$$y := \|u\|_{p,\delta,T},\quad a := K_{50}(\|u_o\|_{p_o} + m(T)^{\varepsilon}U),\quad b := K_{50}m(T)^{\varepsilon}U,\quad \sigma := \gamma\beta.$$

Now (225) is equivalent to $y \leq a + by^{\sigma}$. Hence $y \leq a/(1-\sigma) + b^{1/(1-\sigma)}$ by (203). Thus we get (214) with $K_{49} = K_{50}/(1-\gamma\beta) + K_{50}^{1/(1-\gamma\beta)}$.

Next we prove (215). In the integral equation (45) we estimate the nonlinearity F by the two-sided bound (F1). After splitting the nonlinear term into three factors, (203) and Hölder's inequality (H3) give

$$\begin{aligned}&\|u(t) - S(t)u_o\|_{p_o} \leq \\ &\int_0^t \|S(t-\tau)\|_{p_o,s_1} \|c(.,\tau)\|_{q_1} \|1+|u|(\tau)\|_{r_1}^{\gamma(1-\beta)} \|1+|u|(\tau)\|_p^{\gamma\beta}\, d\tau.\end{aligned} \tag{226}$$

We apply Lemma 6 with s_1, s_2 and the function $f = f(\tau)$ from above and $\delta := 0$, $p := p_o$, $\alpha := \gamma\beta\delta$, $\varepsilon := 0$. By (217)(218) assumption (51) of Lemma 6 is satisfied. Hence (53) implies

$$\|u(t) - S(t)u_o\|_{p_o} \leq K_{51}m(t)^{\varepsilon}P_{s_2}(f)\|1+|u|\|_{p,\delta,t}^{\gamma\beta} \quad \text{for all } t\in(0,T]. \tag{227}$$

Now (224) and (227) imply the assertion (215).

Now we estimate the nonlinearity $\tilde{F} = \tilde{F}(x,t) = F(x,t,u(x,t))$. Again we split into three factors and use Hölder's inequality (H3). Hence we get

$$\begin{aligned}&\int_0^t \|F(.,\tau,u(.,\tau))\|_1\, d\tau \leq \\ &\left[\int_0^t m(\tau)^{-\gamma\beta\delta}\|c(.,\tau)\|_{q_1} \|1+|u|(\tau)\|_{r_1}^{\gamma(1-\beta)}\, d\tau\right] \|1+|u|\|_{p,\delta,t}^{\gamma\beta} \quad \text{for all } t\in(0,T].\end{aligned} \tag{228}$$

Define $\overline{p}\in[1,\infty]$ by $1/\overline{p} = \gamma\beta\delta+\varepsilon$. By (218)(220) and Hölder's inequality (H3), estimate (228) implies

$$\begin{aligned}\int_0^t \|F(.,\tau,u(.,\tau))\|_1\, d\tau &\leq \left[\int_0^t m(\tau)^{1-\varepsilon\overline{p}}d\tau\right]^{1/\overline{p}} U\, \|1+|u|\|_{p,\delta,t}^{\gamma\beta} \\ &\leq K_{52}(t + t^{\varepsilon})U\|1+|u|\|_{p,\delta,t}^{\gamma\beta}\end{aligned} \tag{229}$$

which proves (216).

Finally assume that $u_o \geq 0$, $u \geq 0$, but only the one-sided bound (F3) holds. By Lemma 2(i) the semigroup S(t) is positive. Hence the integral equation (45) and (F3) imply the pointwise estimate

$$0 \leq u(t) \leq S(t)u_o + \int_0^t S(t-\tau)c(.,\tau)(1+u(\tau))\, d\tau. \qquad (230)$$

Then (221) through (225) and finally (214) follow as above.
Thus the Lemma is proved.

Remarks

(i) If $u \geqq 0$, $u_o \geqq 0$ and only the one-sided bound (F3) holds, only estimate (214) remains valid. Estimates (215)(216) can only be shown if the two-sided bound (F1) is available.

(ii) If $\beta > 0$, we have to assume from the beginning that the norm $\|u\|_{p,\delta,T}$ which we want to estimate is indeed finite. Otherwise (225) yields only $\infty \leq \infty$ from which no conclusion can be drawn by feedback. The important point of the Lemma is the detailed structure of estimate (214).

Lemma 20

Let $q_1, q_2 \in [1,\infty]$, $r_1, r_2 \in (0,\infty]$, $\beta \in [0,1]$, $\gamma \in [1,\infty)$, $\varepsilon \in (0,1)$ satisfy

$$n/q_1 + 1/q_2 + \gamma(1-\beta)(n/r_1 + 1/r_2) < 1 - \varepsilon, \qquad (231)$$

$$1/q_1 + \gamma(1-\beta)/r_1 \leq 1, \qquad (232)$$

$$\gamma\beta < 1. \qquad (233)$$

Let $p_o \in [1,\infty]$ be arbitrary.
Let the function F satisfy (FO) and the two-sided local bound (F1) with q_1, q_2, γ from above. If additionally $u_o \geqq 0$, $u \geqq 0$, the one-sided bound (F2) instead of (F1) is sufficient. No Lipschitz condition is needed.

Let the mild solution u of the IBP for initial data $u_o \in L_{p_o}$ on the interval [O,T] satisfy the a priori estimate

$$\|c\|_{q_1,q_2} \||| 1+|u| |||^{\gamma(1-\beta)}_{r_1,r_2,T} = U < \infty. \qquad (213)$$

Furtheron, if $\beta > 0$ assume that

$$\sup\{\|u(t)\|_\infty \mid t \in (0,T]\} < \infty. \quad \text{(See Remark after the proof)}$$

Under these assumptions there exists a constant K_{53} depending on the exponents $p_o, q_1, q_2, r_1, r_2, \beta, \gamma, \varepsilon, N$ and the operator A_o such that

$$\|u(t)\|_\infty \leq K_{53} m(t)^{-n/p_o}[\|u_o\|_{p_o} + m(t)^\varepsilon (U + U^{1/(1-\gamma\beta)})] \qquad (234)$$

for all $t \in (0,T]$.

In the case $\beta = 0$ this means more explicitly that for all $t\in(0,T]$

$$\|u(t)\|_\infty \leq K_{53} m(t)^{-n/p_0}[\|u_0\|_{p_0} + m(t)^\varepsilon \|c\|_{q_1,q_2} \| 1+|u| \|^\gamma_{r_1,r_2,t}] \qquad (235)$$

Proof

If $p_0 = \infty$ or $\beta = 0$, the assumptions (207) through (211) of Lemma 19 are satisfied for $p := \infty, \delta := 0$ and the other exponents as above. Since the assertions (234)(235) are equivalent to the assertion (214) in Lemma 19, the proof is finished.

Now assume that $p_0 \in [1,\infty)$ and $\beta\in(0,1]$. Let $p_0 < p_1 < \ldots < p_k = \infty$ and (δ_i) $i = 1\ldots k$ be the sequences constructed in Lemma 16. Let $\varepsilon\in(0,1)$ be small enough such that (168) as well as (231) hold. Sequentially for $i = 1\ldots k$ we apply Lemma 19 with $p_0 := p_{i-1}$, $p := p_i$, $\delta := \delta_i$, the exponents $q_1, q_2, r_1, r_2, \beta, \gamma, \varepsilon$ from above and initial data $u_0 := u(t)$ for arbitrary $t\in(0,T]$. Hence (214) implies

$$\|u(.,t+.)\|_{p_i,\delta_i,h} \leq K_{49}^{(i)} \|u(t)\|_{p_{i-1}} + K_{49}^{(i)} m(h)^\varepsilon [U + U^{1/(1-\gamma\beta)}] \qquad (236)$$

for all $i = 1\ldots k$, $t\in[0,T)$, $h\in(0,T-t]$.

Choosing $t = ih-h$ we get especially

$$m(h)^{\delta_i} \|u(ih)\|_{p_i} \leq K_{49}^{(i)} \|u(ih-h)\|_{p_{i-1}} + K_{49}^{(i)} m(h)^\varepsilon [U + U^{1/(1-\gamma\beta)}] \qquad (237)$$

for all $i = 1\ldots k$, $h\in(0,T/i]$.

By induction on i this implies

$$m(h)^{(\delta_1+\ldots+\delta_i)} \|u(ih)\|_{p_i} \leq K_{54}^{(i)} \|u_0\|_{p_0} + K_{54}^{(i)} m(h)^\varepsilon [U + U^{1/(1-\gamma\beta)}] \qquad (238)$$

for all $i = 1\ldots k$, $h\in(0,T/i]$.

By (169) we get $\delta_1+\delta_2+\ldots+\delta_k = n/p_0$. Hence (238) for $i = k$ yields the assertion (234). Thus the Lemma is proved.

Remark

The proof shows that the assumption $\sup\{\|u(t)\|_\infty \mid t\in(0,T]\} < \infty$ can be relaxed. Indeed, it is sufficient to assume that

$\sup\{\|u(t)\|_\infty \mid t\in[T_0,T]\} < \infty$ for all $T_0\in(0,T)$ and

$\|u\|_{p_1,\delta_1,T} < \infty$ with p_1, δ_1 from above.

Proof of Theorem 4

We proceed as in the proof of Theorem 3. For initial data $u_o \in L_{p_o}$ there exists a sequence (u_{om}) in the generating set $D \subset L_\infty$ such that according to Definition 3 p.56

$$\lim_{m\to\infty} \|u_{om}-u_o\|_1 = 0 \quad \text{and} \quad \|u_{om}\|_{p_o} \leq \|u_o\|_{p_o} \quad \text{for all } m \in N. \tag{180}$$

Theorem 1 proves the existence of mild solutions u_m of the IBP for initial data u_{om} on the intervals $[0,T_m)$. Let T_m be the maximal existence time from Theorem 1. First we show that $T_m = \infty$ for all $m \in N$.

By the assumptions (193)(194) there exist $\varepsilon\in(0,1)$ and $\beta\in[0,1/\gamma)$ such that the assumptions (231)(232)(233) of Lemma 20 are satisfied. If (202) holds, $\beta = 0$ is sufficient. Estimate (234) implies with U from (242):

$$\|u_m(t)\|_\infty \leq K_{53} m(t)^{-n/p_o}[\|u_o\|_{p_o} + m(t)^{\varepsilon}(U + U^{1/(1-\gamma\beta)})] \tag{240}$$
$$\text{for all } m \in N,\ t\in(0,T_m)\ \{=(0,\infty)\}$$

Hence (58) of Theorem 1 and (240) imply $T_m = \infty$ for all $m \in N$.

Next we show that the sequence u_m converges globally on the time interval $(0,\infty)$. For some $\tilde{T}\in(0,\infty)$ convergence of the restrictions of u_m to the interval $(0,\tilde{T}]$ was already shown by (187) in the proof of Theorem 3. The reasoning is similar to that of p.64. Let $\tilde{p}_o = 1$. There exist $\tilde{p},\tilde{\delta},\tilde{\varepsilon}$ such that the assumptions (147) through (150) of Lemma 15 hold with "$\tilde{\ }$". Choosing the initial time ih for arbitrary $h\in(0,\infty)$ and $i \in\{0\}\cup N$ estimate (153) implies with C from (243) below

$$\|(u_m-u_n)(.,ih+.)\|_{\tilde{p},\tilde{\delta},h} \leq K_{41}\|(u_m-u_n)(ih)\|_1 +$$
$$+ K_{41}\, m(h)^{\tilde{\varepsilon}} C\|(u_m-u_n)(.,ih+.)\|_{\tilde{p},\tilde{\delta},h} \tag{241}$$
$$\text{for all } m,n \in N,\ i \in \{0\}\cup N,\ h\in(0,\infty).$$

The quantities U and C from (240) and (241), respectively, can be estimated by the primary a priori estimate (197):

$$\sup\{\|c\|_{q_1,q_2}|||1+|u_m|\,|||^{\gamma(1-\beta)}_{r_1,r_2,T}\ |\, m \in N,\ T\in(0,\infty)\} = U \leq V \tag{242}$$

$$\sup\{\|c\|_{q_1,q_2}|||1+|u_m|+|u_n|\,|||^{\gamma-1}_{r_1,r_2,T}\ |\, m,n \in N,\ T\in(0,\infty)\} = C \leq V \tag{243}$$

$$\text{with}\quad V = \|c\|_{q_1,q_2}\,(1 + 2G(\|u_o\|_{p_o}))^{\gamma(1-\beta)}. \tag{244}$$

Choose $h \in (0,\infty)$ small enough such that

$$K_{41}\, m(h)^{\varepsilon} V \leq 1/2.$$

Then (241) implies

$$\|(u_m-u_n)(.,ih+.)\|_{p,\delta,h} \leq 2K_{41}\|(u_m-u_n)(ih)\|_1 \quad \text{for all } i \in \{0\} \cup N.$$

Hence the (restrictions of the) functions u_m are a Cauchy sequence in the Banach space $E_{p,\delta,T}$ for all $T \in (0,\infty)$. There exists a function $u \in E_{p,\delta,\infty}$ such that

$$\lim_{m\to\infty} \|u_m - u\|_{p,\delta,T} = 0 \quad \text{for all } T \in (0,\infty).$$

Let $t \in (0,\infty)$ be fixed. By the Theorem of Riesz-Fischer (Segal and Kunze [61] p.97) there exists a subsequence of $u_m(t)$ which converges for almost all $x \in \Omega$. Hence (240) implies that for all $t \in (0,\infty)$

$$\|u(t)\|_\infty \leq K_{53} m(t)^{-n/p_0}[\|u_0\|_{p_0} + m(t)^{\varepsilon}(U + U^{1/(1-\gamma\beta)})]. \tag{245}$$

Hence (200) follows in the special case $p = \infty$, $\delta = n/p_0$. The general case follows as explained on p.66 in the proof of Theorem 3. (244)(245) yield the explicit estimate (201). If (202) holds, let $\beta = 0$ and $\alpha = \gamma$. By the arguments of p.65 the function u is a global mild solution of the IBP for initial data $u_0 \in L_{p_0}$. Thus the Theorem is proved.

As already mentioned in the introduction p.5 we consider two methods for the construction of a priori estimates. The first method - call it "smoothing for short time" - is given in Lemma 22 and 23 below. The second method - call it "bootstrap and feedback" - was already given in Lemma 18 and 19. All these Lemmas do not use a Lipschitz condition Only the bound (F1) for the nonlinearity is important. Even the two-sided bound

(F1) $\quad |F(x,t,u)| \leq c(x,t)(1+|u|)^{\gamma}$ for all $(x,t,u) \in \Omega\times[0,\infty)\times R$

can be replaced by the one-sided bound

(F2) $\quad F(x,t,u) \leq c(x,t)(1 + u)^{\gamma}$ for all $(x,t,u) \in \Omega\times[0,\infty)\times[0,\infty)$

if one has the additional information $u \geq 0$ about the solution u. Thus the assumptions of Lemma 19 and 23 are quite weak, which turns out to be the main advantage in the applications. It is important to make the exponent γ in the assumptions (F1)(F2) as small as possible.

The following Theorem 5 considers the behavior of the solution u in the limit $t \to T_{max}$ for the case $T_{max} < \infty$. The Lipschitz condition is only assumed in the local form (F6) which does not restrict the exponent γ. Of course one can show existence and uniqueness of solutions only in a restricted class of functions and for initial data $u_o \in L_\infty$. This is no severe restriction by the assertion (57) of Theorem 1.

Theorem 5 (Results on the behavior of the solution at a finite maximal existence time, which are available without global Lipschitz condition)

Let the nonlinearity F satisfy (FO), the local Lipschitz condition (F6) and the global two-sided bound (F1) with $q_1, q_2 \in [1,\infty]$, $\gamma \in [1,\infty)$ such that

$$n/q_1 + 1/q_2 < 1 \tag{246}$$

Further restrictions on q_1, q_2, γ are needed in (i) through (v) below. Let u be a mild solution of the initial-boundary value problem (1)(1a)(1b) on the time interval [O,T) such that

$$\sup\{\|u(t)\|_\infty \mid t \in [T_1,T_2]\} < \infty \quad \text{for all } T_1,T_2 \in (0,T). \tag{247}$$

Then the solution u can be extended to a maximal time interval $[0,T_{max})$ such that

$$\sup\{\|u(t)\|_\infty \mid t \in [T_1,T_2]\} < \infty \quad \text{for all } T_1,T_2 \in (0,T_{max}) \tag{248}$$

and the following assertions (i)(ii)(iii)(iv)(v) hold.

If the solution satisfies $u \geq 0$ (on the whole interval $(0,T_{max})$), then (F4) and the one-sided bound (F2) instead of (F1) are sufficient.

(i) If $\gamma = 1$, then $T_{max} = \infty$. (249)

(ii) Assume $\gamma \in (1,\infty)$ and define $p^* \in (0,\infty)$ by

$$n/q_1 + 1/q_2 + (\gamma-1)n/p^* = 1. \tag{250}$$

If $T_{max} < \infty$, the solution u explodes at T_{max} in the following sense:

$$\lim_{t \to T_{max}} \|u(t)\|_p = \infty \quad \text{for all } p \in [1,\infty] \cap (p^*,\infty], \text{ if } T_{max} < \infty. \tag{251}$$

(iii) ("limit case")

Assume that $\gamma\in(1,\infty)$ and p^* given by (250) satisfies $p^*\in(1,\infty)$.

If $T_{max} < \infty$ and the set

$$k(\Pi) = \{u(t)/\|u(t)\|_{p^*} \mid t\in[T_o,T_{max}),\ \|u(t)\|_{p^*} \neq 0\}$$

is precompact in the space $L_{p^*}(\Omega)$ for some (hence all) $T_o \in (0,T_{max})$, then the following holds:

$$\lim_{t\to T_{max}} \|u(t)\|_{p^*} = \infty. \tag{252}$$

(iv) Let $q_1,q_2\in[1,\infty]$, $r_1\in(0,\infty]$, $r_2\in(0,\infty)$, $\gamma\in(1,\infty)$ satisfy

$$n/q_1 + 1/q_2 + (\gamma-1)(n/r_1 + 1/r_2) < 1, \tag{253}$$

$$1/q_1 + (\gamma-1)/r_1 < 1 \quad \text{if} \quad r_1 < \infty. \tag{254}$$

Then the following holds for all $T_o\in(0,T_{max})$:

$$\int_{T_o}^{T_{max}} \|u(t)\|_{r_1}^{r_2}\, dt = \infty \quad \text{if} \quad T_{max} < \infty. \tag{255}$$

(v) Let $q_1,q_2\in[1,\infty]$, $r\in(0,\infty]$, $\rho\in(0,\infty)$, $\gamma\in(1,\infty)$ satisfy

$$n/q_1 + 1/q_2 + (\gamma-1)(n/r + \rho) < 1, \tag{256}$$

$$1/q_1 + (\gamma-1)/r < 1. \tag{257}$$

Then the following holds:

$$\limsup_{t\to T_{max}} (T_{max} - t)^{\rho}\|u(t)\|_r = \infty \quad \text{if} \quad T_{max} < \infty. \tag{258}$$

For the proof we need some Lemmas.

Lemma 21 derives estimates in Hölder spaces from L_∞-estimates. The proof uses the fractional powers A_p^α of the generator A_p of the analytic semigroup $S_p(t)$ in the space $L_p(\Omega)$. Semigroups in the Hölder space as introduced by Kielhöfer [29] can be avoided. From this Lemma we get the compactness needed for the Peano-like argument which is used in Theorem 6 to prove existence of solutions even in the case that no Lipschitz condition holds.

Lemma 22 states the technical basis of the first method to improve a priori estimates.

Lemma 23 constructs L_∞-estimates from L_p-estimates by the first method called "smoothing for short time". This method is different from the "feedback argument" of Lemma 19. The method is based on estimate (68) of Lemma 7, but needs no Lipschitz condition.

Lemma 24 gives the results corresponding to Lemma 23 for the "limit case" $p = p^*$ from (250). The result is much weaker than that of Lemma 23. One has to assume compactness in L_{p^*} and gets no explicit relation between the norms $\|u(t)\|_\infty$ and $\|u(t)\|_{p^*}$ of the solution u.

<u>Lemma 21</u>

Let $q_1, q_2 \in [1,\infty)$, $\nu \in (0,2)$, $\mu, \rho \in (0,1)$ satisfy

$$n/q_1 + 1/q_2 + \nu/2 + \mu < 1, \tag{259}$$

$$n/q_1 + \nu/2 + \mu < \rho. \tag{260}$$

Let the nonlinearity F satisfy (F0) and the local two-sided bound (F4). No Lipschitz condition is needed.

Let u be a mild solution of the IBP for initial data $u_o \in L_\infty$ on the time interval $[0,T]$ or $[0,\infty)$ satisfying for some $U \in (0,\infty)$, respectively,

$$\sup\{\|u(t)\|_\infty \mid t \in (0,T]\} \le U \quad \text{or} \quad \sup\{\|u(t)\|_\infty \mid t \in (0,\infty)\} \le U. \tag{261}$$

Then for all $T_o \in (0,T)$ the (restriction of the) function u is in the Banach space (see p.12 for the Definition) $C^\mu([T_o,T], C^\nu(\bar\Omega))$ or $C^\mu([T_o,\infty), C^\nu(\bar\Omega))$, respectively. There exists a constant K_{55} depending only on the exponents $q_1, q_2, \nu, \mu, \rho, N$ and the operator A_o such that

$$\left.\begin{array}{l} \|u\|_{C^\mu([T_o,T],C^\nu(\bar\Omega))} \\ \text{or} \\ \|u\|_{C^\mu([T_o,\infty),C^\nu(\bar\Omega))} \end{array}\right\} \le K_{55} m(T_o)^{-\rho} \|u_o\|_\infty + K_{55} \|c\|_{q_1,q_2} \Gamma(U), \tag{262}$$

respectively.

<u>Proof</u>

There exist $p \in [q_1,\infty)$, $\alpha, \tilde\beta \in (0,1)$ such that

$$n/q_1 + \alpha - n/p + \mu \le \tilde\beta < \min\{\rho, 1 - 1/q_2\} \text{ and } \nu/2 < \alpha - n/p. \tag{263}$$

By (47) the mild solution satisfies the integral equation

$$u(t_1+t_2) = S(t_2)u(t_1) + \int_0^{t_2} S(t_2-s)F(t_1+s,u(t_1+s))ds \qquad (264)$$
$$\text{for all } t_{1,2} \text{ such that } t_1,t_2,t_1+t_2 \in (0,T].$$

As explained by Henry [24] p.24 one can define fractional powers A_p^α of A_p, the infinitesimal generator of the analytic semigroup $S_p(t)$ in the space $L_p(\Omega)$. By the convention p.24 we simply write $S(t)$ for $S_p(t)$ dropping the index p.

From the equations (45)(47) and the bound (F4) we get the estimates

$$\|A_p^{\alpha+\mu}u(t)\|_p \leq \|A_p^{\alpha+\mu}S(t)\|_{p,q_1}\|u_0\|_\infty + \qquad (265)$$
$$+ \left[\int_0^t \|A_p^{\alpha+\mu}S(t-s)\|_{p,q_1}\|c(.,s)\|_{q_1}\, ds\right]\Gamma(U)$$

and

$$\|A_p^\alpha[u(t_1+t_2)-u(t_1)]\|_p \leq \|A_p^\alpha[S(t_2) - I]u(t_1)\|_p + \qquad (266)$$
$$+ \left[\int_0^{t_2} \|A_p^\alpha S(t_2-s)\|_{p,q_1}\|c(.,t_1+s)\|_{q_1}\, ds\right]\Gamma(U).$$

We have to estimate the right-hand sides of (265)(266). By (30)(39) of Lemma 3 and the semigroup property $S(2t) = S(t)S(t)$ we get

$$\|A_p^\alpha S(2t)\|_{p,q_1} \leq \|A_p^\alpha S(t)\|_{p,p}\|S(t)\|_{p,q_1}$$
$$\leq K_{56}m(t)^{-\alpha}e^{-\lambda t}m(t)^{-(n/q_1-n/p)}e^{-\lambda t}$$

and hence by (263)

$$\|A_p^\alpha S(t)\|_{p,q_1} \leq K_{57}m(t)^{-(\tilde{\beta}-\mu)}e^{-\lambda t} \quad \text{for all } t\in(0,\infty), \qquad (267)$$

$$\|A_p^{\alpha+\mu}S(t)\|_{p,q_1} \leq K_{57}m(t)^{-\tilde{\beta}}e^{-\lambda t} \quad \text{for all } t\in(0,\infty). \qquad (268)$$

Next we estimate the integrals in (265)(266) Define $\bar{q}\in[1,\infty)$ by $1/q_2+1/\bar{q} = 1$. Then (263) implies $\tilde{\beta} < 1/\bar{q}$. The Hölder inequality implies that for all $t\in(0,1)$

$$\int_0^t \|A_p^\alpha S(t-s)\|_{p,q_1}\|c(.,s)\|_{q_1} ds \leq K_{57}\int_0^t m(s)^{-(\tilde{\beta}-\mu)}\|c(.,t-s)\|_{q_1}ds$$
$$\leq K_{57}\left[\int_0^t m(s)^{-(\tilde{\beta}-\mu)\bar{q}}ds\right]^{1/\bar{q}}\|c\|_{q_1,q_2} \leq K_{58}m(t)^{\mu}\|c\|_{q_1,q_2}.$$

By the decomposition into a sum as in the proof of Lemma 6 p.32, the same estimate holds for $t \geq 1$, too. Thus we get

$$\int_0^t \|A_p^\alpha S(t-s)\|_{p,q_1} \|c(.,s)\|_{q_1} ds \leq K_{59} m(t)^\mu \|c\|_{q_1,q_2} \quad \text{and} \tag{269}$$

$$\int_0^t \|A_p^{\alpha+\mu} S(t-s)\|_{p,q_1} \|c(.,s)\|_{q_1} ds \leq K_{60} \|c\|_{q_1,q_2} \quad \text{for all } t\in(0,\infty). \tag{270}$$

Using (268)(270), estimate (265) implies

$$\|A_p^{\alpha+\mu} u(t)\|_p \leq K_{61} m(t)^{-\tilde{\beta}} \|u_0\|_\infty + K_{61} \|c\|_{q_1,q_2} \Gamma(U) \quad \text{for all } t\in(0,T] \tag{271}$$

and $$\|A_p^\alpha u(t)\|_p \leq K_{61} m(t)^{-\tilde{\beta}} \|u_0\|_\infty + K_{61} \|c\|_{q_1,q_2} \Gamma(U) \quad \text{for all } t\in(0,T] \tag{272}$$

by minor modifications.

To estimate the first term in (266), we use (272) and Henry [24], Theorem 1.4.3, p.26 Hence we get

$$\begin{aligned}\|A_p^\alpha [S(t_2)-I]u(t_1)\|_p &\leq \|A_p^{-\mu}[S(t_2)-I]\|_{p,p} |A_p^{\alpha+\mu} u(t_1)|_p \\ &\leq K_{62} m(t_2)^\mu [m(t_1)^{-\tilde{\beta}} \|u_0\|_\infty + \|c\|_{q_1,q_2} \Gamma(U)].\end{aligned} \tag{273}$$

Now (269)(273) imply

$$\|A_p^\alpha [u(t_1+t_2)-u(t_1)]\|_p \leq K_{63} m(t_2)^\mu [m(t_1)^{-\tilde{\beta}} \|u_0\|_\infty + \|c\|_{q_1,q_2} \Gamma(U)] \tag{274}$$

for all $t_{1,2}$ such that $t_1, t_2, t_1+t_2 \in (0,T]$.

By Henry [24], Theorem 1.6.1, p.39, $\nu/2 < \alpha - n/p$ implies

$$C^\nu(\bar{\Omega}) \subset D(A_p^\alpha) \quad \text{and} \quad \|u\|_{C^\nu} \leq K_{64} \|A_p^\alpha u\|_p \quad \text{for all } u \in D(A_p^\alpha). \tag{275}$$

Hence (272)(274) imply that for all $t_{1,2}$ such that $t_1, t_2, t_1+t_2 \in (0,T]$

$$\left.\begin{array}{r} \|u(t_1)\|_{C^\nu} \\ m(t_2)^{-\mu} \|u(t_1+t_2)-u(t_1)\|_{C^\nu} \end{array}\right\} \leq K_{65} [m(t_1)^{-\tilde{\beta}} \|u_0\|_\infty + \|c\|_{q_1,q_2} \Gamma(U)],$$

which proves the assertion (262) of the Lemma since $\tilde{\beta} < \rho$.

Lemma 22

Let $a,b,y\in[0,\infty)$, $\gamma\in(1,\infty)$ and define $b_o,y_o\in(0,\infty)$ by

$$b_o = (\gamma-1)^{(\gamma-1)}(1+a)^{-(\gamma-1)}\gamma^{-\gamma} \quad \text{and} \tag{276}$$

$$y_o = (1+\gamma a)(\gamma-1)^{-1}. \tag{277}$$

(i) There exist two continuously differentiable functions x_1,x_2

x_1: $b\in[0,b_o] \to x_1(b)\in[a,y_o]$ and x_2: $b\in(0,b_o] \to x_2(b)\in[y_o,\infty)$

such that

$$x_1(0) = a, \quad x_1(b_o) = y_o \quad \text{and} \quad dx_1/db > 0 \quad \text{for all } b\in(0,b_o),$$

$$\lim_{b\to 0} x_2(b) = \infty, \quad x_2(b_o) = y_o \quad \text{and} \quad dx_2/db < 0 \quad \text{for all } b\in(0,b_o).$$

Define the sets in the (b,y)-plane

$$\Sigma_> = \{(b,y) \mid b\in[0,b_o),\ y\in(x_1(b),x_2(b))\}$$

$$\partial\Sigma = \{(b,y) \mid b\in[0,b_o],\ y = x_i(b) \ \text{for } i = 1 \text{ or } 2\}$$

$$\Sigma_< = \{(b,y) \mid b\in[0,b_o],\ y\in[0,x_1(b))\cup(x_2(b),\infty)\} \cup (b_o,\infty)\times[0,\infty)$$

Then the following inequalities hold:

$$y > a + b(1+y)^\gamma \quad \text{for all } y \in \Sigma_> \tag{278}$$

$$y = a + b(1+y)^\gamma \quad \text{for all } y \in \partial\Sigma \tag{279}$$

$$y < a + b(1+y)^\gamma \quad \text{for all } y \in \Sigma_< \tag{280}$$

(ii) Let $b = b(t)$ and $x = x(t)$ be two continuous functions

b: $t\in[0,\infty) \to b(t)\in[0,\infty)$ and x: $t\in[0,\infty) \to x(t)\in[0,\infty)$ such that

$b(0) = 0$ and $x(0) = 0$. Then

$$x(t) \leq a + b(t)(1+x(t))^\gamma \ \text{and} \ b(t) < b_o \ \text{for all } t\in(0,\infty) \ \text{imply}$$

$$x(t) < y_o \quad \text{for all } t\in[0,\infty). \tag{281}$$

Proof

Define the function $g: (x,b) \in [0,\infty) \times (0,\infty) \to g(x,b) \in [0,\infty)$ by

$$g(x,b) = a + b(1+x)^{\gamma}.$$

Then $g_x(x,b)$, $g_{xx}(x,b)$, $g_b(x,b) > 0$ for all $(x,b) \in [0,\infty) \times (0,\infty)$. It is elementary to see that the equation

$$g(x,b) = x \tag{282}$$

has exactly two solutions $x_1(b), x_2(b)$ for all $b \in (0,B)$ and no solution for $b \in (B,\infty)$. For $b = B$ equation (282) has exactly one solution $x = x_0$. For B and x_0 we get the equations

$$g(x_0,B) = x_0 \quad \text{and} \quad g_x(x_0,B) = 1. \tag{283}$$

By some computations we get $B = b_0$, $x_0 = y_0$ from (276)(277). Implicit differentiation of

$$g(x_i(b),b) = x_i(b) \quad \text{for all } b \in (0,b_0),\ i = 1,2$$

yields

$$dx_i/db = -(x_i(b)+1)(x_i(b)-a)(x_i(b)-y_0)^{-1} b^{-1} (\gamma-1)^{-1}.$$

Since $x_1(b) < y_0 < x_2(b)$ for all $b \in (0,b_0)$, the assertions about the function $x_i(b)$ follow. Then (278)(280) are clear.

To show (ii) first note that (i) implies

$$(b(t),x(t)) \in \{(b,y) \mid b \in [0,b_0),\ y \in [0,x_1(b)) \cup (x_2(b),\infty)\}.$$

This set consists of two connected components Σ_1 and Σ_2:

$$\Sigma_1 = \{(b,y) \mid b \in [0,b_0),\ y \in [0,x_1(b))\},\ \Sigma_2 = \{(b,y) \mid b \in [0,b_0),\ y \in (x_2(b),\infty)\}.$$

Since the curve $t \in [0,\infty) \to (b(t),x(t))$ is continuous,

$(b(0),x(0)) \in \Sigma_1$ implies

$(b(t),x(t)) \in \Sigma_1$ for all $t \in [0,\infty)$.

Hence $x(t) < x_1(b(t)) < y_0$ for all $t \in [0,\infty)$,

which implies (281). Thus the Lemma is proved.

Lemma 23

Let $p_o, q_1, q_2 \in [1,\infty]$, $\gamma \in (1,\infty)$ satisfy

$$n/q_1 + 1/q_2 + (\gamma-1)n/p_o < 1. \tag{284}$$

Let the nonlinearity F satisfy (FO) and the two-sided global bound (F1) with q_1, q_2, γ from above. Under the additional assumption $u \geqq 0$, the bound (F1) can be replaced by the one-sided bound (F2) and the local two-sided bound (F4). No Lipschitz condition is needed.

Let u be a mild solution of the IBP on the interval $[0,T)$ or $[0,\infty)$. For some $T_o \in (0,T)$ let the solution u satisfy an a priori estimate

$$\sup \{\|u(t)\|_{p_o} \mid t \in (0,T_o]\} \leq V_o < \infty. \tag{285}$$

Furtheron assume

$$\sup \{\|u(t)\|_\infty \mid t \in [T_1,T_2]\} < \infty \quad \text{for all } T_1,T_2 \in (0,T) \text{ or } (0,\infty), \tag{286}$$

respectively.

Then there exists an exponent α depending on p_o, q_1, q_2, γ, N and a constant K_{66} depending on p_o, q_1, q_2, γ, N and the operator A_o such that

$$\sup \{\|u(s)\|_\infty \mid s \in [t, T_o + m(T_o)^\alpha/M]\} \leqq m(t)^{-\alpha} M \quad \text{for all } t \in (0,T_o), \tag{287}$$

$$\|u\|_{p_1,\delta_1,T_o} \leq M \quad \text{for } p_1 \in (p_o,\infty] \text{ specified below, } \delta_1 = n/p_o - n/p_1 \tag{288}$$

with $\qquad M \leq K_{66}(1 + \|c\|_{q_1,q_2} + V_o)^\alpha.$

Proof

By Lemma 12(i) there exists $\varepsilon \in (0,1)$ and finite sequences $p_o < p_1 < \dots < p_k = \infty$ and (δ_i) $i = 1\dots k$ such that (112)(113)(114)(115) hold.

Let t_o, t be such that $0 < t_o \leqq t_o + t < T$. We apply (68) of Lemma 7 to the solution u. Hence we take in (68) $u_m = u_{m+1} = u$, initial data $u_o := u(t_o)$ and exponents $p_o := p_{i-1}$, $p := p_i$, $\delta := \delta_i$ for $i = 1\dots k$ successively. Hence the quantities

$$U_i(t_o,t) = \sup \{m(s)^{\delta_i}\|u(t_o+s)\|_{p_i} \mid s \in [0,t]\} = \|u(.,t_o+.)\|_{p_i,\delta_i,t} \tag{289}$$

satisfy the estimates

$$U_i(t_o,t) \leq K_{30}^{(i)} \|u(t_o)\|_{p_{i-1}} + K_{30}^{(i)} m(t)^{\varepsilon} \|c\|_{q_1,q_2} (1+U_i(t_o,t))^{\gamma} \tag{290}$$
for all $i = 1\ldots k$, $t_o \in (0,T)$, $t \in (0,T-t_o)$.

The assumptions (259)(260) of Lemma 21 are satisfied for some ν,μ,β. Hence by (262) of Lemma 21, the mild solution u is smooth in the sense $u \in C([T_1,T_2],C(\bar{\Omega}))$ for all $T_1,T_2 \in (0,T)$. Hence for all $i = 1\ldots k$, the function $x: t\in[0,T-t_o) \to U_i(t_o,t)$ is continuous and $x(0) = 0$. We apply Lemma 22(ii) with

$$a = K_{30}^{(i)} \|u(t_o)\|_{p_{i-1}}, \quad b(t) = K_{30}^{(i)} m(t)^{\varepsilon} \|c\|_{q_1,q_2}, \quad x(t) = U_i(t_o,t). \tag{291}$$

With these definitions estimate (290) is equivalent to

$$U_i(t_o,t) \leq a + b(t)(1+U_i(t_o,t))^{\gamma} \quad \text{for all } i = 1\ldots k,\ t\in[0,T-t_o). \tag{292}$$

Hence (281) of Lemma 22 shows the following implication:

$$t\in[0,T-t_o),\ b(t) < (\gamma-1)^{(\gamma-1)}\gamma^{-\gamma}(1+a)^{-(\gamma-1)}$$
$$\text{imply} \quad U_i(t_o,t) < (1+\gamma a)/(\gamma-1) \quad \text{for } i = 1\ldots k.$$

After inserting the definitions (291) we get the following implication:

$$i\in\{1\ldots k\},\ t,t_o,t+t_o\in(0,T) \text{ and } K_{67}\, m(t)^{\varepsilon} \|c\|_{q_1,q_2} (1+\|u(t_o)\|_{p_{i-1}})^{\gamma-1} < 1$$
$$\text{imply} \quad (1+\|u(t+t_o)\|_{p_i}) \leq K_{67}\, m(t)^{-\delta_i} (1+\|u(t_o)\|_{p_{i-1}}). \tag{293}$$

We proceed by induction on i. Choose $h\in(0,\infty)$ and a finite sequence $T_o \leq T_1 \leq T_2 \leq \ldots \leq T_k < T$ such that (ih,T_i) is nonvoid for $i = 1\ldots k$. Let

$$V_i = \sup\{\|u(t)\|_{p_i} \mid t\in(ih,T_i)\} \quad \text{for } i = 1\ldots k. \tag{294}$$

By (293) we see that the following implication is true:

$$i\in\{1\ldots k\} \text{ and } K_{67} m(T_i-T_{i-1})^{\varepsilon} \|c\|_{q_1,q_2} (1+V_{i-1})^{\gamma-1} < 1 \quad \text{imply} \tag{295}$$
$$1+V_i \leq K_{67}(1+V_{i-1}) \sup\left\{m(h)^{-\delta_i}, \left[K_{67}\|c\|_{q_1,q_2}(1+V_{i-1})^{\gamma-1}\right]^{\delta_i/\varepsilon}\right\}.$$

For convenience, we give the details of the reasoning leading to (295). To begin with, consider the most simple case that

$$K_{67}\|c\|_{q_1,q_2}(1+V_{i-1})^{\gamma-1} < 1. \tag{296}$$

Applying (293) with $t\in[h,\infty)$, $t_o\in(ih-h,T_{i-1})$ we get

$$1 + \sup\{\|u(s)\|_{p_i} \mid s\in(ih,T)\} < K_{67}m(h)^{-\delta_i}(1+V_{i-1}). \tag{297}$$

Thus the conclusion in (295) is true.

We turn to the case that (296) is violated. Let $t_i\in(0,1]$ be the unique solution of the equation

$$K_{67}\, m(t_i)^{\varepsilon}\|c\|_{q_1,q_2}(1+V_{i-1})^{\gamma-1} = 1. \tag{298}$$

Applying (293) with $t\in(0,t_i)$, $t_o\in(ih-h,T_{i-1})$ such that $t+t_o < T$ yields

$$\begin{aligned}&1 + \sup\{\|u(s)\|_{p_i} \mid s \in (ih-h+t_i,T_{i-1}+t_i)\cap(0,T)\} < \\ &< K_{67}\, m(t_i)^{-\delta_i}(1+V_{i-1}) \le K_{67}(1+V_{i-1})\left[K_{67}\|c\|_{q_1,q_2}(1+V_{i-1})^{\gamma-1}\right]^{\delta_i/\varepsilon}\end{aligned} \tag{299}$$

From the condition in (295) and (298) we get $T_i-T_{i-1} < t_i$. We distinguish the cases (a) $t_i \le h$ and (b) $t_i > h$.

In case (a) we get $(ih-h+t_i,T_{i-1}+t_i) \supset (ih,T_i)$. Hence (299) shows that the conclusion in (295) is true.

In case (b) we apply (293) with $t = h$, $t_o\in(ih-h,T_{i-1})\cap(ih-h,T-h)$. Hence

$$1 + \sup\{\|u(s)\|_{p_i} \mid s \in (ih,T_i+h)\cap(0,T)\} < K_{67}\, m(h)^{-\delta_i}(1+V_{i-1}). \tag{300}$$

Since $[(ih-h+t_i,T_{i-1}+t_i)\cup(ih,T_i+h)]\cap(0,T) \supset (ih,T_i)$, (299)(300) show that the conclusion in (295) is true.

Thus (295) is verified. This statement is exploited by induction on i. One gets the following statements for $i = 1\ldots k$ successively:

There exist exponents α_i depending on $p_i,q_1,q_2,\gamma,\delta_i,\varepsilon,N$ and constants $K_{68}^{(i)}$ depending on $p_i,q_1,q_2,\gamma,\delta_i,\varepsilon,N$ and the operator A_o such that

$$\sup\{\|u(t)\|_{p_i} \mid t\in(ih,T_o)\} \le m(h)^{-\alpha_i}M_i \quad \text{for all } h\in(0,\infty), \text{ where} \tag{301}$$

$$M_i \le K_{68}^{(i)}(1 + V_o + \|c\|_{q_1,q_2})^{\alpha_i}. \tag{302}$$

To verify this successively for $i = 1\ldots k$, one chooses $T_i = T_o$ for all $i = 1\ldots k$. Then the condition in (295) is trivial, hence the conclusion in (295) holds. A simple computation yields (301). In the first step $i = 1$, (301) (not (302)) holds with $\alpha_1 = \delta_1$. Hence (288) follows.

There exist exponents α_i depending on $p_i, q_1, q_2, \gamma, \delta_i, \varepsilon, N$; constants $K_{69}^{(i)}$ depending on $p_i, q_1, q_2, \gamma, \delta_i, \varepsilon, N$ and the operator A_o and numbers $T_i \in (0,T)$ and $M_i \in (0,\infty)$ such that

$$\sup \{\|u(t)\|_{p_i} \mid t \in (ih, T_i)\} \leq m(h)^{-\alpha_i} M_i \quad \text{for all } h \in (0,\infty),$$

$$M_i \leq K_{69}^{(i)} (1 + V_o + \|c\|_{q_1,q_2})^{\alpha_i},$$

$$T_i > T_{i-1} + m(T_o)^{\alpha_i}/M_i.$$

The detailed verification is omitted. For i = k this statement yields the assertion (287). Thus the Lemma is proved.

Lemma 24

Let $p^* \in (1,\infty)$, $q_1 \in [1,\infty]$, $q_2 \in (1,\infty]$, $\gamma \in (1,\infty)$ satisfy

$$n/q_1 + 1/q_2 + (\gamma-1)n/p^* = 1 \quad \text{(with equality sign!)}.$$

Let the nonlinearity F satisfy (FO) and the two-sided global bound (F1) with q_1, q_2, γ from above. Under the additional assumption $u \geq 0$, the bound (F1) can be replaced by the one-sided bound (F2) and the local two-sided bound (F4). No Lipschitz condition is needed.

Let $T \in (0,\infty) \cup \{\infty\}$ and let u be a mild solution of the IBP on the interval $[0,T)$ (possibly $[0,\infty)$) satisfying an a priori estimate

$$\sup \{\|u(t)\|_{p^*} \mid t \in (0,T)\} = V^* < \infty. \tag{303}$$

Furtheron assume that the set

$$k(\Pi) = \{u(t)/\|u(t)\|_{p^*} \mid t \in (0,T),\ \|u(t)\|_{p^*} \neq 0\} \tag{304}$$

is precompact in the space $L_{p^*}(\Omega)$.

Finally assume that

$$\sup \{\|u(t)\|_\infty \mid t \in [T_1,T_2]\} < \infty \quad \text{for all } T_1, T_2 \in (0,T). \tag{305}$$

Then the following estimate restricting the behavior of $\|u(t)\|_\infty$ for $t \to 0$ and $t \to T$ holds for some $\alpha, M \in (0,\infty)$:

$$\sup \{\|u(s)\|_\infty \mid s \in (t,T)\} \leq m(t)^{-\alpha} M \quad \text{for all } t \in (0,T). \tag{306}$$

<u>Proof</u>

By Lemma 12(ii) there exist finite sequences $p^* = p_o < p_1 < \ldots < p_k = \infty$ and (δ_i) $i = 1\ldots k$ such that (112)(113)(114)(115) hold. Especially

$$
\begin{aligned}
n/q_1 + 1/q_2 + (\gamma-1)(\delta_1 + n/p^*) &\leq 1 \\
1/q_1 + \gamma/p_1 &\leq 1 \\
1/q_2 + \gamma\delta_1 &< 1 \\
n/p^* - n/p_1 &= \delta_1 .
\end{aligned}
\tag{307}
$$

We apply Lemma 4 with $p := p^*$, $q := p_1$ and the set

$$\Pi = \{u(t_o) \in L_{p^*}(\Omega) \mid t_o \in (0,T)\}.$$

Hence there exist continuous functions $g = g(t)$, $W = W(t)$ such that

$$\lim_{t\to 0} g(t) = 0 \quad \text{and} \quad m(t) \leq W(t) \leq m(t)^{1/2} \quad \text{for all } t\in(0,\infty) \tag{308}$$

and

$$\|S(t)u(t_o)\|_p \leq K_{13} g(t) m(t)^{-\delta_1} e^{-\lambda t} \|u(t_o)\|_{p^*} \quad \text{for all } t_o\in(0,T), t\in(0,\infty).$$

Let $t_o \in (0,T)$, $t\in[0,T-t_o)$ and define

$$U(t_o,t) = \sup\{W(\tau)^{\delta_1}\|u(t_o+\tau)\|_{p_1} \mid \tau\in[0,t]\} = \|u(.,t_o+.)\|_{W;p_1,\delta_1,t} . \tag{309}$$

We proceed as in the proof of Lemma 8 with $p_o := p^*$, $p := p_1$, $\delta := \delta_1$ and $W = W(t)$ from above. Since assumption (77) of Lemma 8 holds, similarly to (79)(80) we get the following estimate for the mild solution u (corresponding to u_m, u_{m+1} in (79)) and initial data $u_o := u(t_o)$, $t_o \in (0,T)$

$$U(t_o,t) \leq K_{34}\|u(t_o)\|_{p^*} + K_{34} g(t)^{\gamma-1}\|c\|_{q_1,q_2}(1+U(t_o,t))^{\gamma} \tag{310}$$
$$\text{for all } t_o\in(0,T),\ t\in(0,T-t_o).$$

The important point in (309)(310) is that the functions g(t) and W(t) do not depend on $t_o \in (0,T)$.

Furtheron, we proceed as in the proof of Lemma 23 p.84. By Lemma 21 and assumption (305) we get $u \in C([T_1,T_2],C(\overline{\Omega}))$ for all $T_1,T_2 \in (0,T)$.

Hence the function $x: t\in[0,T-t_o) \to U(t_o,t)$ is continuous and $x(0) = 0$. We apply Lemma 22(ii) with

$$a = K_{34}\|u(t_o)\|_{p^*},\quad b(t) = K_{34} g(t)^{\gamma-1}\|c\|_{q_1,q_2},\quad x(t) = U(t_o,t). \tag{311}$$

With these definitions (310) is equivalent to

$$x(t) = a + b(t)(1+x(t))^{\gamma} \quad \text{for all } t\in[0,T-t_o).$$

Hence (281) of Lemma 22 yields the following implication:

$$t\in[0,T-t_o),\ b(t) < (\gamma-1)^{(\gamma-1)}\gamma^{-\gamma}(1+a)^{-(\gamma-1)}$$
$$\text{imply} \quad U(t_o,t) < (1+\gamma a)/(\gamma-1).$$

After inserting (309)(311),$m(t)\leq W(t)$, we get the following implication:

$$t,t_o,t+t_o\in(0,T) \quad \text{and} \quad K_{70}g(t)^{\gamma-1}\|c\|_{q_1,q_2}(1+\|u(t_o)\|_{p^*})^{\gamma-1} < 1 \tag{312}$$
$$\text{imply} \quad (1+\|u(t_o+t)\|_{p_1}) \leq K_{70}m(t)^{-\delta_1}(1+\|u(t_o)\|_{p^*}).$$

By (308) there exists $\bar{\tau}\in(0,T)$ such the condition in (312) holds for all $t\in(0,\bar{\tau})$. Hence (303)(312) imply

$$\sup\{\|u(s)\|_{p_1} \mid s\in(t,T)\} \leq K_{70}m(t)^{-\delta_1}(1+V^*) \quad \text{for all } t\in(0,\bar{\tau}) \quad \text{and}$$

$$\sup\{\|u(s)\|_{p_1} \mid s\in(t,T)\} \leq K_{70}m(t)^{-\delta_1}m(\bar{\tau})^{-\delta_1}(1+V^*) \quad \text{for all } t\in(0,T).$$

Thus (301) holds with $i = 1$, $M_1 = K_{70}m(\bar{\tau})^{-\delta_1}(1+V^*)$. Note that M_1 is not of the type stated in (302), unless the function g satisfies $g(\tau) \leq \tau^{\varepsilon}$.

Now we proceed as in the proof of Lemma 23 proving successively for $i = 2\ldots k$ that (301) holds for some $M_i\in(0,\infty)$ not neccessarily of the type stated in (302). Thus for $i = k$, the assertion (306) follows. Thus the Lemma is proved.

<u>Proof of Theorem 5</u>

We extend the solution u to the maximal time interval $[0,T_{max})$ by the process from the proof of Theorem 1 p.52. Hence we know that

$$\limsup_{t\to T_{max}} \|u(t)\|_{\infty} = \infty \quad \text{if } T_{max} < \infty. \tag{127}$$

(i) Note that in the proof of (152) in Lemma 15 only the global bound (F1) is used, but not the global Lipschitz condition (F3). Hence we may apply (152) in Lemma 15 with $p_o := p := r_1 := r_2 := \infty$, $\delta := 0$, $\gamma = 1$, q_1,q_2 from above satisfying (246) and some $\varepsilon\in(0,1)$.

Let $t_o\in(0,T)$, $t\in[0,T-t_o)$ and define

$$U(t_o,t) = \sup\{\|u(s)\|_{\infty} \mid s\in[t_o,t_o+t]\}. \tag{313}$$

Estimate (152) with initial time t_o, initial data $u_o := u(t_o)$ yields

$$U(t_o,t) \leq K_{41}\|u(t_o)\|_\infty + K_{41}m(t)^\varepsilon\|c\|_{q_1,q_2}(1+U(t_o,t)) \tag{314}$$

$$\text{for all } t_o\in(0,T),\ t\in[0,T-t_o).$$

Choose $h\in(0,T/2)$ small enough such that

$$K_{41}m(h)^\varepsilon\|c\|_{q_1,q_2} \leq 1/2.$$

Then (314) implies

$$U(ih,h) \leq 1 + 2K_{41}\|u(ih)\|_\infty \quad \text{for all } i \in N.$$

Thus recalling (313) we get by induction on i

$$U(ih,h) \leq 1 + 2K_{41} + (2K_{41})^2 + \ldots + (2K_{41})^{i-1} + (2K_{41})^i\|u(h)\|_\infty.$$

Hence (127) implies $T_{max} = \infty$. Note that the solution u grows at most exponentially for $t \to \infty$. Thus (249) is proved.

(ii) Assume that (251) is violated. Then there exists $p_o\in[1,\infty]$ satisfying (284) and a sequence (t_k) such that

$$\lim_{k\to\infty} t_k = T_{max} < \infty \quad \text{and} \quad \sup_{k\in N} \|u(t_k)\|_{p_o} \leq V < \infty. \tag{129}$$

We proceed as in the proof of Lemma 23 p.83,84. Choosing $t_o := t_k$ in (293) we see that the following implication holds:

$$\begin{aligned} &k \in N,\ t\in(0,T_{max}-t_k) \quad \text{and} \quad K_{67}m(t)^\varepsilon\|c\|_{q_1,q_2}(1+\|u(t_k)\|_{p_o})^{\gamma-1} < 1 \\ &\text{imply} \quad 1+\|u(t_k+t)\|_{p_1} \leq K_{67}m(t)^{-\delta_1}(1+\|u(t_k)\|_{p_o}). \end{aligned} \tag{315}$$

Choose $h\in(0,\infty)$ small enough such that

$$K_{67}m(h)^\varepsilon\|c\|_{q_1,q_2}(1+V)^{\gamma-1} < 1. \tag{316}$$

Then (129)(315) imply

$$\sup\{\|u(s)\|_{p_1} \mid s \in \bigcup_{k\in N}(t_k,t_k+h) \cap (0,T_{max})\} \leq K_{67}m(h)^{-\delta_1}(1+V) \tag{317}$$

and

$$\sup\{\|u(s)\|_{p_1} \mid s \in (T_o,T_{max})\} < \infty \quad \text{for some } T_o\in(0,T_{max}). \tag{318}$$

Hence Lemma 23 (indeed just the successive use of the argument above for the chain of exponents $p_o < p_1 < \ldots < p_k = \infty$) yields

$$\sup \{\|u(s)\|_\infty \mid s\in(T_o,T_{max})\} < \infty \quad \text{for some } T_o\in(0,T_{max}). \tag{319}$$

By (127) estimate (319) can only hold if $T_{max} = \infty$. Thus we get a contradiction. Hence (251) must hold.

(iii) Assume that the assertion in (iii) does not hold. Then there exists a sequence (t_k) such that

$$\lim_{k\to\infty} t_k = T_{max} < \infty \quad \text{and} \quad \sup_{k\in N} \|u(t_k)\|_{p*} \leq V < \infty. \tag{320}$$

We proceed as in the proof of Lemma 24 p.87,88. Choosing $t_o := t_k$ in (312) we get the following implication:

$$\begin{aligned} &k \in N,\ t\in(0,T_{max}-t_k) \quad \text{and} \quad K_{70}g(t)^{\gamma-1}\|c\|_{q_1,q_2}(1+\|u(t_k)\|_{p*})^{\gamma-1} < 1 \\ &\text{imply} \quad 1+\|u(t_k+t)\|_{p_1} \leq K_{70}m(t)^{-\delta_1}(1+\|u(t_k)\|_{p*}). \end{aligned} \tag{321}$$

Choose $h\in(0,\infty)$ small enough such that

$$K_{70}g(h)^{\gamma-1}\|c\|_{q_1,q_2}(1+V)^{\gamma-1} < 1.$$

Then (320)(321) imply

$$\sup \{\|u(s)\|_{p_1} \mid s \in \bigcup_{k\in N}(t_k,t_k+h) \cap (0,T_{max})\} \leq K_{70}m(h)^{-\delta_1}(1+V)$$

and

$$\sup \{\|u(s)\|_{p_1} \mid s \in (T_o,T_{max})\} < \infty \quad \text{for some } T_o\in(0,T_{max}). \tag{322}$$

Hence Lemma 23 (succesive estimates of L_{p_i}-norms for $i = 2\ldots k$) implies (319). By (127) this implies $T_{max} = \infty$, since the L_∞-norm "explodes" at a finite T_{max}. Thus we have arrived at a contradiction. Hence (iii) must hold.

(iv) Assume that the assertion in (iv) is violated. Then $T_{max} < \infty$ and for some $T_o\in(0,T_{max})$ we get

$$\int_{T_o}^{T_{max}} \|u(t)\|_{r_1}^{r_2}\, dt \leq V < \infty. \tag{323}$$

We apply Lemma 20. By assumption (253)(254) there exist $\beta\in[0,1/\gamma)$ and $\varepsilon\in(0,1)$ such that the assumptions (231)(232)(233) of Lemma 20 hold. We take T_o as initial time and initial data $u_o := u(T_o)$. Assumption (213) is satisfied for all $T < T_{max}-T_o$ since

$$\|c\|_{q_1,q_2} \||1+ u(.,T_o+.)|\|^{\gamma(1-\beta)}_{r_1,r_2,T_{max}-T_o} \leq U < \infty.$$

Hence (234) with $p_o = \infty$ shows that

$$\|u(t)\|_\infty \leq K_{53}\|u(T_o)\|_\infty + K_{53}(U + U^{1/(1-\gamma\beta)}) \quad \text{for all } t\in(T_o,T_{max}).$$

By (127) this estimate can only hold for $T_{max} = \infty$. Thus we get a contradiction. Hence (iv) must hold.

(v) Assume that (v) is false. Then $T_{max} < \infty$ and for some $T_o\in(0,T_{max})$

$$\|u(t)\|_r \leq M(T_{max} - t)^{-\rho} \quad \text{for all } t\in(T_o,T_{max}). \tag{324}$$

Let $r_1 := r$ and choose $r_2\in(1/\rho,\infty)$. Then (324) implies (323). On the other hand, assumptions (253)(254) of (iv) are satisfied. Hence (255) holds. Thus we get (323) and (255) which contradict one another. Hence (v) must hold.

Thus the Theorem is proved.

<u>Theorem 6</u> (Global existence and uniform a priori estimates in the case without global Lipschitz condition)

(i) Let $q_1,q_2\in[1,\infty]$, $r_1,r_2\in(0,\infty]$, $\gamma\in[1,\infty)$ satisfy

$$n/q_1 + 1/q_2 + (\gamma-1)(n/r_1 + 1/r_2) < 1, \tag{325}$$

$$1/q_1 + (\gamma-1)/r_1 < 1 \quad \text{if} \quad r_1 < \infty. \tag{326}$$

Let $p_o\in[1,\infty]$ be arbitrary.

Let the nonlinearity F satisfy (F0), the two-sided global bound (F1) and the local Lipschitz condition (F6) with q_1,q_2,γ from above.

Assume that there exists a generating set D (see Definition 3 p.56) and an increasing function G: $u\in[0,\infty) \to G(u)\in[0,\infty)$ such that every mild solution u of the initial-boundary value problem (1)(1a)(1b) for initial data $u_o \in D$ on any time interval $[0,T]$ satisfies an a priori estimate

$$\|u\|_{r_1,r_2,T} \leq G(\|u_o\|_{p_o}). \tag{327}$$

(Note that we assume this estimate for every $T > 0$ for which it makes sense, not only for some T given from the beginning as in Theorem 3).

Then for every initial data $u_o \in L_{p_o}(\Omega)$ the initial-boundary value problem (1)(1a)(1b) has a global mild solution u on the time interval $[0,\infty)$. Furthermore this solution satisfies

$$\lim_{t\to 0} \|u(t) - S(t)u_o\|_{p_o} = 0 \tag{328}$$

$$\lim_{t\to 0} \|u(t) - u_o\|_{p_o} = 0 \quad \text{if assumption (A) p. 20 holds} \tag{329}$$

$$\sup_{0<t<\infty} m(t)^{\delta}\|u(t)\|_p < \infty \quad \text{for all } p\in[p_o,\infty],\ \delta = n/p_o - n/p. \tag{330}$$

There exist exponents $\alpha\in(0,\infty)$, $\varepsilon\in(0,1)$ depending only on the exponents $p_o,q_1,q_2,r_1,r_2,\gamma,N$ and a constant K_{48} depending on these exponents and on the operator A_o such that

$$\|u(t)\|_\infty \leq K_{48}m(t)^{-n/p_o}[\|u_o\|_{p_o} + m(t)^{\varepsilon}(\Phi+\Phi^{\alpha})], \text{ where} \tag{331}$$

$$\Phi = (1 + \|c\|_{q_1,q_2})(1 + G(\|u_o\|_{p_o})).$$

If the initial data satisfy $u_o \geqq 0$ and the nonlinearity F satisfies the positivity **assumption** (F7) then the solution u satisfies $u \geqq 0$.

(ii) Let $p_o,q_1,q_2\in[1,\infty]$, $\gamma\in[1,\infty)$ satisfy

$$n/q_1 + 1/q_2 + (\gamma-1)/n/p_o < 1. \tag{332}$$

Let the nonlinearity F satisfy (F0), the two-sided global bound (F1) and the local Lipschitz condition (F6) with q_1,q_2,γ from above.

Assume that there exists a generating set D (see Definition 3 p.56) and an increasing function G: $u\in[0,\infty) \to G(u)\in[0,\infty)$ such that every mild solution u of the initial-boundary value problem (1)(1a)(1b) for initial data $u_o \in D$ on any time interval $[0,T]$ satisfies an a priori estimate

$$\sup\{\|u(t)\|_{p_o} \mid t\in(0,T]\} \leq G(\|u_o\|_{p_o}). \tag{333}$$

(Note that we assume this estimate for every $T > 0$ for which it makes sense, not only for some T given from the beginning).

Then for every initial data $u_o \in L_{p_o}(\Omega)$ the initial-boundary value problem (1)(1a)(1b) has a global mild solution u on the time interval $[0,\infty)$. Furthermore this solution satisfies (328) and (329) from above.

There exists an exponent $\alpha \in (0,\infty)$ depending on the exponents p_0, q_1, q_2, γ, N and a constant K_{66} depending on these exponents and on the operator A_0 such that

$$\|u(t)\|_\infty \leq K_{66} m(t)^{-\alpha}[1+\|c\|_{q_1,q_2}+G(\|u_0\|_{p_0})]^\alpha \quad \text{for all } t\in(0,\infty). \tag{334}$$

Furthermore, if $1/q_1+(\gamma-1)/p_0 \leq 1$ or $1/q_2+\gamma n/p_0 < 1$, (335a) or (335b) then (330) holds.

If the initial data satisfy $u_0 \geqq 0$ and the nonlinearity F satisfies the positivity assumption (F7) then the solution u satisfies $u \geqq 0$.

<u>Remark</u>

In the case that $u_o \geqq 0$, $u \geqq 0$ it would be desirable to replace the two-sided bound (F1) by the one-sided bound (F2) as e.g. in Theorem 5 or other assertions above. This seems to be impossible. See p.97 for further comments.

<u>Proof of Theorem 6</u>

(i) There exist $\varepsilon\in(0,1)$ and $\beta\in[0,1/\gamma)$ such that the assumptions (231) (232)(233) of Lemma 20 are satisfied. Furtheron, by Lemma 16 there exist $p\in(p_0,\infty]$, $\delta\in(0,1)$ (they are called p_1, δ_1 in Lemma 16) such that the assumptions (207) through (211) of Lemma 19 are satisfied.

For initial data $u_0 \in L_{p_0}(\Omega)$ there exists a sequence (u_{om}) in the generating set $D \subset L_\infty$ such that according to Definition 3 p.56

$$\lim_{m\to\infty} \|u_{om}-u_0\|_1 = 0 \quad \text{and} \quad \|u_{om}\|_{p_0} \leq \|u_0\|_{p_0} \quad \text{for all } m \in N. \tag{336}$$

Theorem 1 proves the existence of mild solutions u_m of the IBP for initial data u_{om} on the intervals $[0,T_m)$. Let T_m be the maximal existence time from Theorem 1. First we show that $T_m = \infty$ for all $m \in N$.

We apply Lemma 19 and 20. By the estimates (214)(215)(229) we get

$$\|u_m(t)-S(t)u_{om}\|_{p_0} \leq K_{71} m(t)^\varepsilon U[1+\|u_0\|_{p_0}+(U+U^{1/(1-\gamma\beta)})] \tag{337}$$
$$\text{for all } m \in N,\ t\in(0,T_m),$$

$$\int_0^t \|F(s,u_m(s))\|_1\, ds \leq K_{71}(t+t^\varepsilon)U[1+\|u_0\|_{p_0}+(U+U^{1/(1-\gamma\beta)})] \tag{338}$$
$$\text{for all } m \in N,\ t\in(0,T_m).$$

By (234) of Lemma 20 we get

$$\|u_m(t)\|_\infty \leq K_{53} m(t)^{-n/p_0}[\|u_0\|_{p_0} + m(t)^\varepsilon(U+U^{1/(1-\gamma\beta)})] \tag{339}$$
$$\text{for all } m \in N,\ t\in(0,T_m).$$

Here U is defined by (213) and can be estimated by assumption (327):

$$U = \sup \{\|c\|_{q_1,q_2} \|\!|1+|u_m|\|\!|^{\gamma(1-\beta)}_{r_1,r_2,T} \mid m \in N,\ T\in(0,T_m)\} \leq V \quad \text{where} \tag{340}$$

$$V = \|c\|_{q_1,q_2} (1 + 2G(\|u_o\|_{p_o}))^{\gamma(1-\beta)}. \tag{341}$$

Estimate (339) and the explosion property (127) in Theorem 1 (or, alternatively (337) and (58)) imply $T_m = \infty$. Hence u_m are global mild solutions for initial data u_{om}.

We cannot prove that the sequence u_m converges, for no appropriate global Lipschitz condition is assumed as was the case in Theorem 3 and 4. Instead we use compactness to extract a converging subsequence from the sequence (u_m). Let t_o, T_o such that $0 < t_o < T_o < \infty$ be arbitrary and $\nu, \mu\in(0,2)$ such that assumption

$$n/q_1 + 1/q_2 + \nu/2 + \mu < 1 \tag{259}$$

of Lemma 21 holds. Taking initial time t_o and initial data $u_o := u_m(t_o)$, estimate (262) of Lemma 21 implies that for all $m \in N$, $0 < t_o < T_o < \infty$:

$$\|u_m\|_{C^\mu([T_o,\infty),C^\nu(\overline{\Omega}))} \leq K_{55} m(T_o-t_o)^{-\rho} \|u_m(t_o)\|_\infty + $$
$$+ K_{55}\|c\|_{q_1,q_2} \Gamma[\sup\{\|u_m(t)\|_\infty \mid t\in(t_o,\infty)\}].$$

Hence estimate (339) implies the bound uniform in $m \in N$:

$$\sup_{m\in N} \|u_m\|_{C^\mu([T_o,\infty),C^\nu(\overline{\Omega}))} \leq M \quad \text{for all } T_o\in(0,\infty), \nu,\mu\in(0,2) \text{ such that } n/q_1+1/q_2+\nu/2+\mu < 1. \tag{342}$$

Here $M\in(0,\infty)$ depends on $\nu, \mu, N, p_o, T_o, \|u_o\|_{p_o}, \|c\|_{q_1,q_2}, q_1, q_2, \gamma, K_{53}, K_{55}$ and the functions G, Γ. Naturally, in (342) u_m denotes the restriction of the solution $u_m = u_m(.,t)$ to the time interval $[T_o,\infty)$.

By the Theorem of Ascoli the embedding $C^{\tilde{\mu}}([T,\infty),C^{\tilde{\nu}}(\overline{\Omega})) \subset C^\mu([T,\infty),C^\nu(\overline{\Omega}))$ is compact if $\tilde{\nu} > \nu$, $\tilde{\mu} > \mu$. Hence by applying a standard diagonal argument with respect to T_o, ν, μ, (342) shows that there exists a subsequence (u_{m_j}) of (u_m) and a function $u: (x,t)\in\overline{\Omega}\times(0,\infty) \to u(x,t)\in R$ such that

$$\lim_{j\to\infty} \|u_{m_j} - u\|_{C^\mu([T_o,\infty),C^\nu(\overline{\Omega}))} = 0 \quad \text{for all } T_o\in(0,\infty),\ \nu,\mu\in(0,2) \text{ such that } n/q_1+1/q_2+\nu/2+\mu < 1. \tag{343}$$

For all $t\in(0,\infty)$, the sequences $u_j(t)$, $F(t,u_j(t))$ and by (31) Lemma 3(ii) and (336) the sequence $S(t)u_{oj}$, too, converge in the norm of $C^\nu(\overline{\Omega})$.

(We simply write u_j for the subsequence u_{m_j}) Taking the limit $j \to \infty$, estimates (337)(338)(339)(340)(341) imply that for some $W\in(0,\infty)$:

$$\|u(t) - S(t)u_0\|_{p_0} \leq K_{71} m(t)^{\varepsilon} W \qquad \text{for all } t\in(0,\infty), \tag{344}$$

$$\int_0^t \|F(s,u(s))\|_1 \, ds \leq K_{71}(t + t^{\varepsilon})W \qquad \text{for all } t\in(0,\infty), \tag{345}$$

$$\|u(t)\|_{\infty} \leq K_{71} m(t)^{-n/p_0} W \quad \text{for all } t\in(0,\infty). \tag{346}$$

Now (344) proves (328) and via Lemma 2(iv) we get (329), too.
To show (330), notice that (344) and (346) yield the assertion (330) for the special cases $p = p_0$, $\delta = 0$ and $p = \infty$, $\delta = n/p_0$, respectively. Hence the interpolation inequality

$$\|u\|_p \leq \|u\|_{p_0}^{p_0/p} \, \|u\|_{\infty}^{1-p_0/p} \qquad \text{for all } p\in[p_0,\infty],\ u \in L_{\infty}(\Omega)$$

implies that (330) holds for arbitrary $p\in[p_0,\infty]$, too.
Finally (331) follows from (339)(340)(341) again by taking the limit $j \to \infty$, $u_j \to u$.

We have to show that the function u is indeed a global mild solution of the IBP for initial data u_0 on the interval $[0,\infty)$. Since the functions u_j are mild solutions for initial data u_{0j}, they satisfy

$$u_j(t) = S(t)u_{0j} + \int_0^t S(t-s)F(s,u_j(s))ds \quad \text{for all } t\in(0,\infty), j \in N. \tag{347}$$

We split the integral at some $h\in(0,t)$ and get

$$u_j(t) = S(t)u_{0j} + \int_h^t S(t-s)F(s,u_j(s))ds + R_j(h) \tag{348}$$

with the "rest"

$$R_j(h) = S(t-h)[S(h)u_{0j} - u_j(h)]. \tag{349}$$

The limit for $j \to \infty$ exists for all terms, if one takes e.g. the metric of $C^{\nu}(\bar{\Omega})$. Hence we get

$$u(t) = S(t)u_0 + \int_h^t S(t-s)F(s,u(s))ds + R(h) \tag{350}$$

with the "rest"

$$R(h) = S(t-h)[S(h)u_0 - u(h)]. \tag{351}$$

As a second step, we take the limit $h \to 0$. Let $\alpha > \nu/2 + n/p_o$. By (344) and Lemma 3(i)(ii) we estimate the rest (351):

$$\|R(h)\|_{C^\nu} \le \|S(t-h)\|_{C^\nu, p_o} K_{71} m(h)^{\varepsilon} W \le K_{72} m(t-h)^{-\alpha} m(h)^{\varepsilon} W \quad \text{for all } h \in (0,t). \tag{352}$$

Hence we get

$$\lim_{h\to 0} \|R(h)\|_{C^\nu(\bar{\Omega})} = 0. \tag{353}$$

Thus the function u satisfies the integral equation (45). By estimate (345) the integral in (45) is absolutely converging in $L_1(\Omega)$. Hence u is a mild solution of the IBP for initial data u_o on the time interval $[0,\infty)$ in the sense of Definition 1 p.29.

(ii) As in (i) we approximate the initial data $u_o \in L_{p_o}(\Omega)$ by a sequence (u_{om}) in the generating set $D \subset L_\infty$ such that (336) holds. Let u_m be the mild solutions of the IBP for initial data u_{om} on the maximal time intervals $[0,T_m)$. To show $T_m = \infty$ for all $m \in N$, we apply Lemma 23.

As in the proof of Lemma 23 we begin by choosing $\varepsilon \in (0,1)$ and finite sequences $p_o < p_1 < \ldots < p_k = \infty$ and (δ_i) $i = 1\ldots k$ such that (112)(113)(114)(115) hold. By assertions (287)(288) of Lemma 23 there exists an exponent α depending on p_o, q_1, q_2, γ, N such that

$$\|u_m(t)\|_\infty \le m(t)^{-\alpha} M \quad \text{for all } m \in N,\ t \in (0,T_m), \tag{354}$$

$$\|u_m(t)\|_{p_1} \le m(t)^{-\delta_1} M \quad \text{for all } m \in N,\ t \in (0,T_m), \tag{355}$$

where $M \in (0,\infty)$ can be estimated by (285) and the assumption (333):

$$M \le K_{66}(1 + \|c\|_{q_1,q_2} + G(\|u_o\|_{p_o}))^{\alpha}. \tag{356}$$

Since assumption (63) of Lemma 7 is satisfied with $p = p_1$, $\delta = \delta_1$, we can use estimates (73)(76) which are proved p.37,38. Note that the Lipschitz condition is not needed to this end. Hence we get

$$\int_0^t \|F(s,u_m(s))\|_1 \, ds \le K_{32}(1 + t)\|c\|_{q_1,q_2} \|1+|u_m|\|^{\gamma}_{p_1,\delta_1,t}, \tag{357}$$

$$\|u_m(t) - S(t)u_{om}\|_{p_o} \le K_{33} m(t)^{\varepsilon} \|c\|_{q_1,q_2} \|1+|u_m|\|^{\gamma}_{p_1,\delta_1,t} \quad \text{for all } m \in N,\ t \in (0,T_m). \tag{358}$$

Estimate (354) and the explosion property (127) in Theorem 1 imply that $T_m = \infty$. Hence u_m are global mild solutions for all $m \in N$.

Furtheron, we proceed as in (i). A subsequence (u_{m_j}) can be extracted from the sequence (u_m) which converges to a function u in the sense of (343). Hence (357)(355) imply (345), (358)(355) imply (344) and assertion (328)(239). Clearly (354)(355)(356) imply assertion

$$\|u(t)\|_\infty \leq K_{66} m(t)^{-\alpha}[1 + \|c\|_{q_1,q_2} + G(\|u_o\|_{p_o})]^\alpha \text{ for all } t\in(0,\infty), \quad (334)$$

$$\|u(t)\|_{p_1} \leq K_{66} m(t)^{-\delta_1}[1 + \|c\|_{q_1,q_2} + G(\|u_o\|_{p_o})]^\alpha \text{ for all } t\in(0,\infty). \quad (359)$$

It remains to show (330) under the assumption (335a) or (335b). If (335a) holds we can apply part (i) with $r_1 := p_o$, $r_2 := \infty$. If (335b) holds, we can choose $p_1 = \infty$, $\delta_1 = n/p_o$. Hence (359) implies (330). Since (344)(345) hold, the arguments of p. 95 show that the function u is a mild solution of the IBP for initial data u_o on the time interval $[0,\infty)$.

Let the nonlinearity F satisfy the positivity assumption (F7) and take initial data $u_o \doteq 0$. Then the approximating sequence (u_{om}) can be chosen such that $u_{om} \doteq 0$ for all $m \in N$. Hence Lemma 10 shows for the mild solutions $u_m \doteq 0$ for all $m \in N$. Hence (343) implies $u \doteq 0$.

Thus the Theorem is proved.

Remark

It seems to be impossible to replace the two-sided bound (F1) by the one-sided bound (F2) in the case that $u_o \doteq 0$ and (F7) holds. Clearly one can choose approximating data $u_{om} \doteq 0$ and get approximating solutions $u_m \doteq 0$ converging to $u \doteq 0$. Estimates (339)(354)(355) can be shown. All this can be managed by assuming only the one-sided bound (F2). The difficulty comes in, if one tries to get (337)(358), which assure that the initial data u_{om} are assumed for $t \to 0$ uniformly in $m \in N$. These estimates can only be proved if one assumes the two-sided bound (F1).

In the case of sublinear nonlinearities F which satisfy the bounds

$$|F(x,t,u)| \leq c(x,t)(1+|u|) \quad \text{for all } (x,t)\in\Omega\times[0,\infty),\ u \in R,$$

$$|F(x,t,u)-F(x,t,v)| \leq c(x,t)|u - v| \quad \text{for all } (x,t)\in\Omega\times[0,\infty),\ u,v \in R$$

i.e. (F1)(F3) with $\gamma = 1$, we can improve the results above. Indeed, it is sufficient to assume a weaker primary a priori estimate than $|||u|||_{r_1,r_2,T} < \infty$ as in Theorem 3,4,5 and 6 above.

The following Definition is needed to formulate weak primary estimates.

Definition 4

Let $H: u\in[0,\infty) \to H(u)\in[1,\infty)$ be any continuous function such that

$$\lim_{u\to\infty} H(u) = \infty. \tag{359}$$

Let $T\in(0,\infty)\cup\{\infty\}$ and let $\mathfrak{M}$ be the set of all measurable functions

$u: (x,t)\in\Omega\times(0,T) \to u(x,t)\in R.$

We define a functional

$\Phi_T: u\in\mathfrak{M} \to \Phi_T(u)\in[0,\infty]$ by setting

$$\Phi_T(u) = \sup\left\{\int_{t_1}^{t_2}\int_\Omega H(|u(x,s)|)dxds \;\middle|\; t_1,t_2\in(0,T),\ 0\le t_2-t_1\le 1\right\}. \tag{360}$$

To begin with, we prove some Lemmas:

Lemma 25 constructs a function $\tilde{H}$ from H, which is more convenient in the primary a priori estimate, especially for initial data $u_o \notin L_\infty$.

Lemma 26 applies the "feedback"-argument explained in the introduction p.6 in the case of weak primary a priori estimates and nonlinearities F which satisfy (F1) with $\gamma = 1$.

Lemma 25

Let $r_1\in(0,\infty)$ and let $H: u\in[0,\infty) \to H(u)\in[1,\infty)$ be a continuous function satisfying (359).

Then there exists a continuous function $\tilde{H}: u\in[0,\infty) \to \tilde{H}(u)\in[1,\infty)$ with the following properties:

$$1 \le \tilde{H}(u) \le H(u) \quad \text{for all } u\in[0,\infty), \tag{361}$$

$$\lim_{u\to\infty} \tilde{H}(u) = \infty. \tag{362}$$

Furtheron, the function $\tilde{h} = \tilde{h}(u) := u\tilde{H}(u)^{-1/r_1}$ satisfies

$$\tilde{h}(u) \le \tilde{h}(\mu u) \le \mu\tilde{h}(u) \quad \text{for all } u\in[0,\infty),\ \mu\in[1,\infty). \tag{363}$$

Proof

Define a continuous function $h: u\in[0,\infty) \to h(u)\in(0,\infty)$ by setting

$$h(u) := uH(u)^{-1/r_1}. \tag{364}$$

Clearly the function $h(u)$ satisfies

$$h(0) = 0 , \quad 0 < h(u) \leq u \quad \text{for all } u \in (0,\infty), \tag{365}$$

$$\lim_{u\to\infty} h(u)/u = 0. \tag{366}$$

Define the function $\tilde{h}: u \in [0,\infty) \to \tilde{h}(u) \in [0,\infty)$ by setting

$$\tilde{h}(u) = \sup \{\lambda h(a)+(1-\lambda)h(b) \mid a,b \in [0,\infty),\ \lambda \in [0,1],\ u=\lambda a+(1-\lambda)b\}. \tag{367}$$

By some computation one checks that $\tilde{h}$ has the following properties:

(a) $\tilde{h}(0) = 0 , \quad 0 < h(u) \leq \tilde{h}(u)$ for all $u \in (0,\infty)$

(b) $\lambda\tilde{h}(a)+(1-\lambda)\tilde{h}(b) \leq \tilde{h}(\lambda a+(1-\lambda)b)$ for all $a,b \in [0,\infty)$, $\lambda \in [0,1]$
i.e. the function $\tilde{h}$ is concave.

(c) If the function $g: u \in [0,\infty) \to g(u) \in [0,\infty)$ is concave and $h \leq g$ then $\tilde{h} \leq g$.

The properties (a)(b)(c) characterize the function $\tilde{h}$ uniquely. Indeed, $\tilde{h}$ is the least concave function larger than h. We derive further properties of the function $\tilde{h}$ from (365)(366).

$$\tilde{h}(0) = 0 , \quad 0 < h(u) \leq \tilde{h}(u) \leq u \quad \text{for all } u \in (0,\infty) \tag{368}$$

follows from (a) and furtheron by (c) with $g(u) = u$ and (365).
Now assertion (363) is a straightforward consequence of (368) and the concavity of the function $\tilde{h}$.

It remains to show (362). By (366), for any $\varepsilon > 0$ there exists $M(\varepsilon)$ such that

$$h(u) \leq \varepsilon u + M(\varepsilon) \quad \text{for all } u \in [0,\infty).$$

Hence property (c) implies

$$\tilde{h}(u) \leq \varepsilon u + M(\varepsilon) \quad \text{for all } u \in [0,\infty)$$

and

$$\limsup_{u\to\infty} \tilde{h}(u)/u \leq \varepsilon.$$

Since $\varepsilon > 0$ is arbitrary this implies

$$\lim_{u\to\infty} \tilde{h}(u)/u = 0. \tag{369}$$

Define the function $\tilde{H}$ by setting $\tilde{h}(u) = u\tilde{H}(u)^{-1/r_1}$. Then (368) implies the assertion (361) and (369) implies (362). Thus the Lemma is proved.

Lemma 26

Let $p_o \in [1,\infty]$, $q_1,q_2 \in (1,\infty]$, $\varepsilon \in (0,1)$ satisfy

$$n/q_1 + 1/q_2 + n/p_o < 1 - \varepsilon. \tag{370}$$

Let the nonlinearity F satisfy (F0) and the two-sided bound (F1) with q_1,q_2 from above and $\gamma = 1$. If one knows additionally that $u_o \geqq 0$, $u \geqq 0$, the two-sided bound (F1) can be replaced by the one-sided bound (F2). No Lipschitz condition is needed.

Assume that u is a mild solution of the IBP for initial data $u_o \in L_{p_o}$ on the time interval [0,T] satisfying an a priori estimate

$$\Phi_T(u) < \infty \tag{371}$$

for some functional Φ_T specified in Definition 4 p.98. Furtheron, let

$$\sup \{\|u(t)\|_\infty \mid t \in (0,T]\} < \infty. \tag{372}$$

Then there exists a constant $M \in (0,\infty)$ depending on the exponents $p_o,q_1,q_1,q_2,\varepsilon,N$, the operator A_o, the norms $\|u_o\|_{p_o}, \|c\|_{q_1,q_2}$ and the function H and the functional $\Phi_T(u)$ in Definition 4 (not on T) such that

$$\|u(t)\|_\infty \leqq m(t)^{-n/p_o} M \qquad \text{for all } t \in (0,T], \tag{373}$$

$$\|u(t) - S(t)u_o\|_{p_o} \leqq m(t)^{\varepsilon} M \quad \text{for all } t \in (0,T]. \tag{374}$$

Remark

Note that M does not depend on the supremem occuring in (372), otherwise the Lemma would be trivial. Indeed, assumption (372) is only needed to excluded that (383) simply yields $\infty = \infty$, from which no conclusion can be drawn by feedback.

Proof

There exist $r_1,r_2,r \in (1,\infty)$ such that

$$n/q_1 + 1/q_2 + n/r_1 + 1/r_2 < 1 - \varepsilon, \tag{375}$$

$$1/q_1 + 1/r_1 \leqq 1, \tag{376}$$

$$1/r_1 + 1/r = 1/r_2. \tag{377}$$

We apply Lemma 19 with $p_o,q_1,q_2,r_1,r_2,\varepsilon$ above, $p:=\infty$, $\beta:=0$, $\gamma:=1, \delta:=n/p_o$.

By (370)(375)(376) the assumptions (207) through (211) of Lemma 19 are satisfied. Hence assertions (213)(214)(215) imply

$$\|u\|_{\infty,\delta,t} \le K_{49}\|u_0\|_{p_0} + 2K_{49}\|c\|_{q_1,q_2}|||1+|u|\,|||_{r_1,r_2,t} \qquad \text{for all} \quad (378)$$

$$t\in(0,T].$$

$$\|u(t) - S(t)u_0\|_{p_0} \le K_{49}m(t)^{\varepsilon}\|c\|_{q_1,q_2}|||1+|u|\,|||_{r_1,r_2,t} \qquad (379)$$

Define $U:=\|u\|_{\infty,\delta,T}$ with $\delta = n/p_0$ and let $\tilde{h},\tilde{H}$ be the functions constructed in Lemma 25. For all $t_1,t_2\in(0,T]$ with $0\le t_2-t_1\le 1$ we estimate:

$$\left[\int_{t_1}^{t_2} \|u(t)\|_{r_1}^{r_2}\,dt\right]^{1/r_2} = \left[\int_{t_1}^{t_2}\left(\int_\Omega \tilde{h}(|u(x,t)|)^{r_1}\tilde{H}(|u(x,t)|)dx\right)^{r_2/r_1}dt\right]^{1/r_2}$$

$$\le \left[\int_{t_1}^{t_2} \tilde{h}(\|u(t)\|_\infty)^{r_2}\left(\int_\Omega \tilde{H}(|u(x,t)|)dx\right)^{r_2/r_1}dt\right]^{1/r_2}$$

$$\le \tilde{h}(U)\left[\int_{t_1}^{t_2} m(t)^{-\delta r_2}\left(\int_\Omega \tilde{H}(|u(x,t)|)dx\right)^{r_2/r_1}dt\right]^{1/r_2} \qquad (380)$$

$$\le \tilde{h}(U)\left[\int_{t_1}^{t_2} m(t)^{-\delta r}\,dt\right]^{1/r}\left[\int_{t_1}^{t_2}\left(\int_\Omega \tilde{H}(|u(x,t)|)dx\right)dt\right]^{1/r_1} \qquad (381)$$

$$\le K_{73}\,\tilde{h}(U)\,\Phi_T(u)^{1/r_1} = K_{73}\,\Phi_T(u)^{1/r_1}\,U\,\tilde{H}(U)^{-1/r_1}$$

Here (380) follows from (363), (381) follows from (377) and Hölder's inequality. Altogether we get

$$|||u|||_{r_1,r_2,T} \le K_{73}\,\Phi_T(u)^{1/r_1}\,U\,\tilde{H}(U)^{-1/r_1}. \qquad (382)$$

Hence (378)(382) imply

$$U \le K_{49}\|u_0\|_{p_0} + K_{74}\|c\|_{q_1,q_2}[1 + \Phi_T(u)^{1/r_1}\,U\,\tilde{H}(U)^{-1/r_1}] \qquad (383)$$

Since

$$\lim_{U\to\infty} \tilde{H}(U) = \infty$$

by (362), the right-hand side of (383) grows less than the left-hand side for $U \to \infty$. Thus (383) contains a "feedback". Since $U < \infty$ by assumption (372), estimate (373) follows. Finally (379)(382) imply the assertion (374). Thus the Lemma is proved.

<u>Theorem 7</u> (Stronger results for sublinear equations using only weak primary a priori estimates)

Let $p_o \in [1,\infty]$, $q_1, q_2 \in (1,\infty]$ satisfy

$$n/q_1 + 1/q_2 + n/p_o < 1. \tag{384}$$

Let the nonlinearity F satisfy (FO), the two-sided global bound

$$|F(x,t,u)| \leq c(x,t)(1+|u|) \quad \text{for all } (x,t) \in \Omega \times [0,\infty),\ u \in R,$$

which is just (F1) with $\gamma = 1$, finally the local Lipschitz condition (F6) with q_1, q_2 from above.

Assume that there exists a generating set D (see Definition 3 p.56) and an increasing function G: $u \in [0,\infty) \to G(u) \in [0,\infty)$ and a functional Φ_∞ as specified by Definition 4 p.98 such that every mild solution u of the initial-boundary value problem (1)(1a)(1b) for initial data $u_o \in D$ on any time interval [O,T] satisfies an a priori estimate

$$\Phi_T(u) \leq G(\|u_o\|_{p_o}). \tag{385}$$

(Note that we assume this estimate for every T > O for which it makes sense, not only for some T given from the beginning).

Then for every initial data $u_o \in L_{p_o}(\Omega)$ the initial-boundary value problem (1)(1a)(1b) has a global mild solution on the time interval $[0,\infty)$. Furthermore this solution satisfies

$$\lim_{t\to 0} \|u(t) - S(t)u_o\|_{p_o} = 0 \tag{386}$$

$$\lim_{t\to 0} \|u(t) - u_o\|_{p_o} = 0 \quad \text{if assumption (A) p.20 holds} \tag{387}$$

$$\sup_{0<t<\infty} m(t)^{\delta} \|u(t)\|_p < \infty \quad \text{for all } p \in [p_o,\infty],\ \delta = n/p_o - n/p \tag{388}$$

If the initial data satisfy $u_o \geq 0$ and the nonlinearity satisfies the positivity assumption (F7), then the solution satisfies $u \geq 0$.

<u>Remarks</u>

(i) We cannot prove uniqueness, unless the global Lipschitz condition (F2) with $\gamma = 1$ is assumed.

(ii) It seems to be impossible to handle the case of positive solutions $u \geq 0$ by the one-sided bound (F2). See Remarks p.93 and 97.

Corollary of Theorem 7

We relaxe the assumptions of Theorem 7:
Let $T_o \in (0,\infty)$ be arbitrary and assume that the primary a priori estimate (385) holds only for $T \leq T_o$. The other assumptions of Theorem 7 remain unchanged.

Then for every initial data $u_o \in L_{p_o}(\Omega)$ the initial-boundary value problem (1)(1a)(1b) has a global mild solution u on the time interval $[0,\infty)$. Furthermore this solution satisfies (386)(387). Instead of (388) the following is true: There exists $M \in (0,\infty)$ such that

$$\sup_{0<s<t} m(s)^{\delta} \|u(s)\|_p \leq M e^{Mt} \quad \text{for all } p \in [p_o,\infty],\ t \in (0,\infty), \quad \delta = n/p_o - n/p. \tag{389}$$

Proof

For initial data $u_o \in L_{p_o}(\Omega)$ there exists a sequence (u_{om}) in the generating set $D \subset L_\infty$ such that according to Definition 3 p.56

$$\lim_{m\to\infty} \|u_{om} - u_o\|_1 = 0 \quad \text{and} \quad \|u_{om}\|_{p_o} \leq \|u_o\|_{p_o} \quad \text{for all } m \in N. \tag{390}$$

Theorem 1 proves the existence of mild solutions u_m of the IBP for initial data u_{om} on the intervals $[0,T_m)$. Let T_m be the maximal existence time from Theorem 1. First we show that $T_m = \infty$ for all $m \in N$.

We apply Lemma 26. Choose $\varepsilon \in (0,1)$ such that (370) holds. Assertions (373)(374) of Lemma 26 show that there exists $M \in (0,\infty)$ such that

$$\|u_m(t)\|_\infty \leq m(t)^{-n/p_o} M \qquad \text{for all } m \in N,\ t \in (0,T_m), \tag{391}$$

$$\|u_m(t) - S(t)u_{om}\|_{p_o} \leq m(t)^{\varepsilon} M \quad \text{for all } m \in N,\ t \in (0,T_m). \tag{392}$$

Estimate (391) and the explosion property (127) in Theorem 1 imply $T_m = \infty$. Hence, for all $m \in N$, the functions u_m are global mild solutions for initial data u_{om}. Now we proceed as in the proof of Theorem 6 p.94. A subsequence of (u_m) converges to the function u in the sense of (343). One shows (344)(345)(346) for this function u and proves that u is indeed the global mild solution of the IBP for initial data u_o on the interval $[0,\infty)$. Thus the Theorem is proved.

Under the relaxed assumptions of the Corollary we get only a solution of the IBP on the interval $[0,T_o]$. By Theorem 5(i) this solution can be extended on $[0,\infty)$. The proof of Theorem 5(i) p.89 shows that the function u grows at most exponentially for $t \to \infty$. Hence (389) follows.

Part II Some Examples of Reaction-Diffusion Systems Arising in Applications

In this part, for some reaction-diffusion systems occuring in applications, global existence of solutions is shown and in some cases the asymptotic behavior is investigated.

As a first step, primary - possibly very rough - a priori estimates are derived. This step exploites the structure of the special system considered. It may be trivial - as in example 4, chemical reactions - or quite involved - as in example 1, the Gierer-Meinhardt model, or example 5, a nuclear reactor problem. In this step, explicit calculations, Lyapunov functionals, the maximum principle and comparison arguments are used. To justify these tools, we have to work with smooth initial data and classical solutions. For convenience of the reader, the basic Theorems needed are reviewed briefly.

To derive uniform a priori estimates from the primary estimate, the methods and results of part I - especially Lemma 19, 20, 23 and Theorem 5, 6 - are used. By the assumptions needed there, restrictions of the space dimension may enter. It turns out that many modifications exploiting the structure of the special example are necessary to get optimal results.

Although it is not our main interest, we try to handle irregular initial data, too, by approximating them by smooth data. Thus some proofs get quite lengthy. By the compactness-method of Theorem 6,7 we prove existence of solutions for irregular initial data even in cases not covered by Theorem 1 or 2 of part I. To this end, the primary a priori estimates must be of the structure like (327) or (385). Hence one must keep track of the quantities - as e.g. L_1- or L_2-norms or other functionals of the initial data, constants in the equations, etc. - which enter into the calculations leading to the primary a priori estimate. Thus e.g. $C^{\nu}(\bar{\Omega})$- or $C^{2+\alpha}(\bar{\Omega})$-norms of the initial data must not occur in the primary a priori estimate, although smoothness of the solution can indeed be assumed. For systems containing ordinary differential equations - as example 3, the FitzHugh-Nagumo system, or example 5, a nuclear reactor model, indeed do - we are restricted to the method of Theorem 3,4. Hence the dependence of the solution on the initial data must be studied in detail, typically the solutions depend Lipschitz continuously on their initial data.

Some specific remarks for each example should be made:

1. The Gierer-Meinhardt model was studied numerically by the inventors in view to different applications in developmental biology. There is little work from the analytical point of view. A local bifurcation analysis including stability considerations is done for example by Takagi [63]. A global bifurcation analysis is contained in Mimura and Nishiura [41]. Hadeler, Rothe and Vogt [21] prove existence of equilibria by degree theory. All these papers are only concerned with the equilibrium states and their linearized stability properties. The author does not know any publication concerned with the global behavior of the time dependent system. This aspect is considered here, indeed we prove existence of globally bounded solutions. One cannot expect existence of a Lyapunov functional or global stability. Indeed, there can occur Hopf bifurcations and multiple steady states - in the one-dimensional case this is evident by symmetry breaking.

2. The Brusselator was one of the first examples for an open chemical system far from equilibrium which exhibits interesting phenomena (see Prigogine and Glansdorff [48] for some background). There exists a vaste literature about the Brusselator. For example, Auchmuty and Nicolis [5] give a local bifurcation analysis and Herschkowitz-Kaufmann [25] has done many numerical studies. Existence of a globally bounded solution is not proved in their work. Some weak results concerning global existence under additional hypotheses were given by Lange [23].

3. The FitzHugh-Nagumo system is the possibly most simplified model for the spread of a nerve pulse along the axon (see e.g. the article of FitzHugh in Schwan [60] for an excellent review of the background). Carpenter [7] has proved the existence of a pulse solution by topological methods. Evans and Feroe [12] give a thourough study of the local stability. Rauch and Smoller [51] show existence of a globally bounded solution for the time dependent equations by the elementary method of contracting rectangles. To this end, they have to impose strong restrictions on the nonlinear term. These restrictions can be relaxed by the methods of this monograph. Furtheron we use a Lyapunov functional suggested by Lopes [34] to give conditions under which every solution of the time dependent equations decays to zero, i.e. the nerve is dead.

4. Chemical reactions governed by the mass action law were indeed the starting point of this work. To prove global existence of bounded solutions turns out to be nontrivial, although this seems to be clear

intuitively. An a priori estimate of the L_1-norm is evident by the conservation of mass. In case of space dimension $N = 1$ one can apply Theorem 1 (in the obvious vector-valued version) directly. This result can be much improved. Indeed, modifying the bootstrap and feedback argument in a manner convenient for the system yields global bounds up to space $N = 5$. The negative entropy turns out to be a Lyapunov functional which is useful to study the asymptotic behavior. W. Ebel [11] investigated this system as a model for transport through membranes and proposed more realistic boundary conditions.

5. The nuclear reactor model considered was introduced by Kastenberg and Chambré [26]. Local stability was shown by Mottoni and Tesei [42]. In this monograph we prove that the solutions of the time dependent equations are bounded globally in space and time and that the equilibrium state is globally stable. The main tool is a Lyapunov functional. The construction of this functional uses an idea of Kawasaki and Teramoto [27]. The results of part I show compactness of the trajectory in a Hölder space and even compactness of space-time dependent "pieces" of the trajectory in a Hölder space of space-time functions. This setting is well-known in the theory of time-lag equations, but seems to be new in this context. As an advantage one gets differential equations for the space-time dependent functions in the limit set for $t \to \infty$. Indeed, only all these methods together are sufficient to prove global stability of the equilibrium state.

6. The Volterra-Lotka model describes the ecological interaction of prey and predator (see e.g. May [37] for the background in theoretical ecology). Indeed existence of global solutions was first studied for the predator-prey system by Alikakos [1,2]. Our method is in some way simpler than that of Alikakos (see the remarks in the introduction p.7). An L_1-estimate follows by Jensen's inequality quite easily from the well-known Lyapunov functional. This L_1-estimate is the primary a priori estimate that we improve by the methods of part I. The food pyramide structure introduced by Williams and Chow [69] gives a natural order in which the components are considered: The results of part I are applied at first to the equation for the prey, then to the equation for the predator. After global bounds in space and time are established, it is quite easy to show that the solution converges to a homogeneous equilibrium for $t \to \infty$.
The degenerate case that only one species diffuses turns out to be rather delicate. The asymptotic behavior is complicated, because there exists a continuum of equilibrium states.

A more complete study of the degenerate case is started in Rothe [57] and will be continued. In this monograph, we consider the degenerate case only for Dirichlet boundary conditions and positive initial data. Then we get convergence to the homogeneous equilibrium for $t \to \infty$. We can prove L_1-convergence of the diffusing component, but only weak convergence of the nondiffusing component. It remains an open problem how to prove strong convergence of the nondiffusing component.

Finally, we consider some slightly generalized Volterra-Lotka systems. We construct a Lyapunov functional as a sum of two parts, each of which depends only on one component. For the class of systems considered, the solution remains globally bounded and approaches the homogeneous equilibrium for $t \to \infty$.

Naturally, part II containes only some typical examples. Indeed, these examples gave the motivation to write this monograph. Surely the methods used here could be applied to many other systems. It remains an open question how to characterize a reasonably large class of systems, for which the solutions remain globally uniformly bounded. By the opinion of the author, one has to study the structure of the system in phase space to this end. One has to take care not to get totally absorbed by the functional analytic part of the work built up in part I at the beginning and by many other authors.

Review of Standard Theorems

Let the positive integer N denote the space dimension. For simpler notation we define $n = N/2$. Thus the half-integer n is half of the space dimension. Denote by $x = (x_1, x_2, \ldots, x_N)$ a generic point in R^N. Let Ω be a bounded domain in R^N whose boundary $\partial\Omega$ is an $(N-1)$-dimensional $C^{2+\alpha}$-manifold for some $\alpha \in (0,1)$ such that Ω lies locally on one side of $\partial\Omega$. Denote by $(x,t) = (x_1, x_2, \ldots, x_N, t)$ a generic point in $R^N \times R$. In the following "x" is the space coordinate and "t" is the time.

Let $m \in N$ be a positive integer and let $u = (u_1, u_2, \ldots, u_m)$ be a generic point in the phase space R^m. In the following "u_ι" for $\iota = 1 \ldots m$ are some physically measurable quantities.

Let $d_\iota = 0$ for $\iota = 1 \ldots r$, $d_\iota > 0$ for $\iota = r+1 \ldots m$ ($r = 0$ may occur) and $D = \mathrm{diag}(d_1, d_2, \ldots, d_N)$ the diagonal matrix of diffusion coefficients.

Let the function $u_0 = (u_{01}, \ldots, u_{0m})(x_1, \ldots, x_N)$;

$$u_0: \qquad x \in \Omega \to u_0(x) \in R^m$$

model the initial data.

Let the function $F = (F_1, \ldots, F_m)(x_1, \ldots, x_N, t, u_1, \ldots, u_m)$;

$$F: (x,t,u) \in \bar{\Omega} \times [0,\infty) \times R^m \to F(x,t,u) \in R^m$$

be the nonlinearity, modelling e.g. chemical reactions.

We look for solutions $u = (u_1, \ldots, u_m)(x_1, \ldots x_N, t)$;

$$u: \qquad (x,t) \in \bar{\Omega} \times [0,\infty) \to u(x,t) \in R^m$$

of the reaction-diffusion system (1)(2)(3) below. The time interval $[0,\infty)$ must be restricted and extended and cannot be definitively prescribed at this starting point.

We consider the reaction-diffusion system

$$u_t - D\Delta u + Du = F(x,t,u) \quad \text{for all } x \in \Omega,\ t > 0 \tag{1}$$

together with the boundary conditions (2) and the initial condition (3).

Let $0 \leq r \leq s \leq m$. By $(n_1(x),\ldots,n_N(x))$ we denote the outer normal unit vector at the boundary point $x \in \partial\Omega$. Let

$$\partial/\partial n = \sum_{j=1}^{N} n_j(x)\,\partial/\partial x_j$$

be the derivative in the outer normal direction.

For $\iota = s+1\ldots m$ we need functions $b_\iota \in C^{1+\alpha}(\partial\Omega)$.
Altogether, the regularity assumption for the boundary conditions are:

$$\partial\Omega \in C^{2+\alpha} \quad \text{and} \quad b_\iota \in C^{1+\alpha}(\partial\Omega) \quad \text{for } \iota = s+1\ldots m, \tag{BO}$$

$$b_\iota(x) \geq 0 \qquad \text{for } \iota = s+1\ldots m, \text{ all } x \in \partial\Omega. \tag{B1}$$

We assume the following boundary conditions (2a)(2b)(2c):

$$d_\iota = 0 \quad \text{for } \iota = 1\ldots r; \tag{2a}$$

$$u_\iota(x,t) = 0 \quad \text{for } \iota = r+1\ldots s,\ x \in \partial\Omega,\ t > 0; \tag{2b}$$

$$b_\iota(x)u_\iota(x,t) + \partial u_\iota/\partial n(x,t) = 0 \quad \text{for } \iota = s+1\ldots m,\ x \in \partial\Omega,\ t > 0. \tag{2c}$$

Finally we assume the initial conditions

$$u(x,0) = u_0(x) \quad \text{for all } x \in \Omega. \tag{3}$$

Note that components which do not diffuse as well as different types of boundary conditions for the diffusing components may occur.

The construction of solutions of the initial-boundary value problem (1)(2)(3) begins with bounded mild solutions and then turns to classical solutions. The regularity assumptions (M) and (S) are adapted to these cases, respectively.

$$u_0 \in L_\infty(\Omega, R^m) \tag{MO}$$

The function $F(.,.u)$: $(x,t) \in \bar{\Omega}\times[0,\infty) \to F(x,t,u) \in R^m$ is measurable in (x,t) for all $u \in R^m$. (MFO)

For every bounded set $B \subset \bar{\Omega}\times[0,\infty)\times R^m$, there exists a constant $L(B)$ such that

$$|F(x,t,u)| \leq L(B) \qquad \text{for all } (x,t,u) \in B, \tag{MF1}$$

$$|F(x,t,u)-F(x,t,v)| \leq L(B)\,|u-v| \quad \text{for all } (x,t,u),(x,t,v) \in B. \tag{MF2}$$

$$u_{0\iota} \in C^{\alpha}(\overline{\Omega}) \quad \text{for } \iota = 1\ldots r; \qquad u_{0\iota} \in C^{2+\alpha}(\overline{\Omega}) \quad \text{for } \iota = r+1\ldots m. \tag{S0}$$

$$\left.\begin{aligned} u_{0\iota}(x) &= 0 \\ -d_\iota \Delta u_{0\iota}(x) &= F_\iota(x,0,u_0(x)) \end{aligned}\right\} \quad \text{for } \iota = r+1\ldots s,\ x\in\partial\Omega; \tag{S1b}$$

$$b_\iota(x)u_{0\iota}(x)+\partial u_{0\iota}/\partial n(x) = 0 \qquad \text{for } \iota = s+1\ldots m,\ x\in\partial\Omega. \tag{S1c}$$

For every bounded set $B \subset \overline{\Omega}\times[0,\infty)\times R^m$,
there exists a constant $L(B)$ such that

$$|F(x,t,u)| \le L(B) \quad \text{for all } (x,t,u) \in B, \tag{SF1}$$

$$|F(x,t,u)-F(y,s,v)| \le L(B)[|x-y|^{\alpha}+|t-s|^{\alpha/2}+|u-v|] \quad \text{for all } (x,t,u),(y,s,v) \in B. \tag{SF2}$$

Assumptions (S1b) mean that the initial data $u_{0\iota}$ satisfy the boundary condition (2b) and the differential equation (1) for $\iota = r+1\ldots s$. Assumption (S1c) means that the initial data $u_{0\iota}$ satisfy the boundary condition (2c) for $\iota = s+1\ldots m$.
Comparing with Theorems 5.2 and 5.3 in Ladyzenskaja [32] p.320 with $l := \alpha$ from above, we see that (S1b) and (S1c) are just the compatibility conditions of order $[l/2]+1 = 1$ and $[(l+1)/2] = 0$, respectively, which are needed in Theorems 5.2 and 5.3 of [32], respectively.

Definition 1

Initial data u_0 satisfying (S0)(S1) are called regular.
The set of regular initial data is denoted by $\mathcal{R}$.

As a first step in the study of the initial-boundary value problem (1)(2)(3), we consider the linear system, i.e. (1)(2)(3) with $F \equiv 0$. The solution $u = u(x,t)$ of this system for initial data $u_0 \in L_\infty(\Omega,R^m)$ defines a formal semigroup $P(t)$ in the space $L_\infty(\Omega,R^m)$ by setting

$$u(x,t) = (P(t)u_0)(x) \quad \text{for all } x \in \Omega,\ t\in[0,\infty).$$

Since the linear system is decoupled, the operator $P(t)$ acts componentwise. Let for all $t\in[0,\infty)$

$$P_\iota(t) = I \quad \text{for } \iota = 1\ldots r, \qquad P_\iota(t) = S_\iota(d_\iota t)\big|_{L_\infty} \quad \text{for } \iota = r+1\ldots m,$$

where I denotes the identity operator and $S_\iota(t)$ the semigroup defined p.15,24 generated by the operator $-\Delta+1$ and the boundary conditions (2b) or (2c) for $\iota = r+1...s$ or $\iota = s+1...m$, respectively.
Thus the semigroup P(t) is given by

$$P(t) = (P_1(t), P_2(t), \dots, P_N(t)) \quad \text{for all } t \in [0,\infty).$$

By the Remark p.23, P(t) is not a strongly continuous semigroup in L_∞.
In the Banach space $L_\infty(\Omega,R^m)$ we use the norm

$$\|u\|_\infty = \max_{1\le\iota\le m} \|u_\iota\|_\infty \quad \text{for all } u \in L_\infty(\Omega,R^m).$$

The maximum principle (see p.123 below or Lemma 2(ii) p.19) implies

$$\|P(t)u_0\|_\infty \le \|u_0\|_\infty \quad \text{for all } u_0 \in L_\infty(\Omega,R^m),\ t\in[0,\infty). \tag{4}$$

Definition 2

Let $T\in(0,\infty)\cup\{\infty\}$. A $E_{\infty,0,T}$-mild solution of the initial-boundary value problem (1)(2)(3) for initial data $u_0 \in L_\infty(\Omega,R^m)$ on the time interval $[0,T)$ is a measurable function $u: (x,t)\in\Omega\times(0,T) \to u(x,t)\in R^m$ satisfying

$$u(.,t) \in L_\infty(\Omega) \quad\text{and}\quad \sup\{\|u(.,s)\|_\infty | s\in(0,t)\} < \infty \quad \text{for all } t\in(0,T), \tag{5}$$

$$u(.,t) = P(t)u_0 + \int_0^t P(t-s)F(.,s,u(.,s))\,ds \quad \text{for all } t\in(0,T), \tag{6}$$

where the integral is an absolutely converging Bochner integral in $L_\infty(\Omega)$.

This notion of $E_{\infty,0,T}$-mild solution is because of (5) much more restrictive than that of mild solution from Definition 1 p.29 or L_r-mild solution from Definition 2 p.49. Let $u(s)=u(.,s)$, $F(s,u(s))=F(.,s,u(.,s))$.

Theorem 1 (Existence of mild and classical solutions)

Assume that (BO)(B1)(MO)(MFO)(MF1)(MF2) hold. Then (i)(ii)(ii) are true.

(i) For each initial function $u_0 \in L_\infty(\Omega,R^m)$, there exists $T\in(0,\infty]$ such that the initial-boundary value problem (1)(2)(3) has a unique $E_{\infty,0,T}$-mild solution on the interval $[0,T)$.

(ii) Consider the existence time T as a functional of the initial data $u_0 \in L_\infty(\Omega,R^m)$. Then this functional $T = T(u_0)$ satisfies

$$\inf \{T(u_o) \mid u_o \in L_\infty(\Omega,R^m),\ \|u_o\|_\infty \leq U_o\} > 0 \quad \text{for all } U_o \in [0,\infty). \tag{7}$$

(iii) The existence time $T \in (0,\infty) \cup \{\infty\}$ can be chosen maximal (i.e. (i) does not hold for a larger time). In that case we let $T = T_{max}$ and get

$$\lim_{t \to T_{max}} \|u(t)\|_\infty = \infty \quad \text{if } T_{max} < \infty . \tag{8}$$

Assume (SO)(S1) i.e. regular initial data $u_o \in \mathcal{R}$ and (SF1)(SF2).

(iv) Then the mild solution u is smooth. With $u(.,0) = u_o$ we have

$$u_\iota \in C^{\alpha,1+\alpha/2}(\bar{\Omega}\times[0,T]) \qquad \text{for } \iota = 1\ldots r, \text{ all } T \in (0,T_{max}), \tag{9}$$

$$u_\iota \in C^{2+\alpha,1+\alpha/2}(\bar{\Omega}\times[0,T]) \quad \text{for } \iota = r+1\ldots m, \text{ all } T \in (0,T_{max}) \tag{10}$$

(see e.g. Ladyzenskaja [32] p.8 for the definition of these Hölder spaces). Furtheron, the function u satisfies the differential equation (1) on the closed domain, i.e. for all $(x,t) \in \bar{\Omega}\times[0,T_{max})$, and the boundary condition (2) for all $(x,t) \in \partial\Omega\times[0,T_{max})$ including $x \in \partial\Omega$ or $t = 0$.

Remark

This Theorem seems to be well-known (see e.g. Amann [3], Henry [24], Kielhöfer [29]). Nevertheless we could not find it in the literature in the form stated here. Usually the explosion property (8) is only stated for some norm involving smoothness, but not the L_∞-norm. Also systems with nondiffusing components or third type boundary conditions are less common in the literature. For these reasons and the convenience of the reader we give a proof, which turns out to be rather simple.

Proof

Let $U_o \in [0,\infty)$ be given. Choose $U \in (U_o,\infty)$, $T_o \in (0,\infty)$ arbitrary and define the bounded set $B := \bar{\Omega}\times[0,T_o]\times[-U,U]^m$. Let $L(B) \in (0,\infty)$ be the constant appearing in (MF1)(MF2) and choose $T \in (0,T_o]$ such that

$$U_o + e^{L(B)T} - 1 \leq U. \tag{11}$$

(This can only understand a mathematician knowing the following proof!) For initial data $u_o \in L_\infty(\Omega,R^m)$ with $\|u_o\|_\infty \leq U_o$ we define the following Picard-Lindelöf sequence $u^{(i)}$ in $L_\infty(\bar{\Omega}\times[0,T],R^m)$:

$$u^{(1)}(t) = P(t)u_o \ ; \quad u^{(i+1)}(t) = P(t)u_o + \int_0^t P(t-s)F(s,u^{(i)}(s))\,ds \tag{12}$$

for $i \in N$, $t \in [0,T]$.

For all $i \in N$ define the functions $\eta_i : t\in[0,T] \to \eta_i(t)\in[0,\infty)$ by setting

$$\eta_i(t) = \|(u^{(i+1)}-u^{(i)})(t)\|_\infty. \tag{13}$$

We show by induction on i that the following holds for all $t\in[0,T]$, $i\in N$:

$$\|u^{(i)}(t)\|_\infty \leq U, \tag{14}$$

$$\eta_i(t) \leq L(B)\int_0^t \eta_{i-1}(s)ds \quad \text{for } i > 1, \tag{15}$$

$$\eta_i(t) \leq (L(B)t)^i/i!, \tag{16}$$

$$\sum_{1\leq j\leq i} \eta_j(t) \leq e^{L(B)t} - 1. \tag{17}$$

The well-known computations to check this involve (MF1)(MF2)(4)(11)(12). Begin with the case i = 1. By (4)(12) we get

$$\|u^{(1)}(t)\|_\infty \leq \|P(t)u_0\|_\infty \leq \|u_0\|_\infty \leq U_0 < U.$$

By (MF1)(12)(13) we get

$$\eta_1(t) \leq \int_0^t \|F(s,u^{(1)}(s))\|_\infty ds \leq L(B)t.$$

Suppose that (14)(15)(16)(17) is already shown for some $i-1 \in N$. We have to prove these assertions for $i\in N$. By (MF2)(4)(12)(13) we get

$$\begin{aligned}\eta_i(t) &\leq \int_0^t \|P(t-s)[F(s,u^{(i)}(s))-F(s,u^{(i-1)}(s))]\|_\infty\, ds \\ &\leq L(B)\int_0^t \eta_{i-1}(s)ds.\end{aligned}$$

Hence (15) is proved. (16) follows by explicit integration and (17) by the series for the exponential function. It remains to show (14). By (11)(13)(14)(17) we get

$$\begin{aligned}\|u^{(i)}(t)\|_\infty &\leq \|u^{(1)}(t)\|_\infty + \sum_{1\leq j<i} \eta_j(t) \\ &\leq U_0 + e^{L(B)t} - 1 \\ &\leq U \quad \text{proving (14).}\end{aligned}$$

Hence (14)(15)(16)(17) hold for all $i \in N$.

There exists a function $u \in L_\infty(\bar{\Omega}\times[0,T], R^m)$ such that

$$\sup\{\|(u^{(i)}-u)(t)\|_\infty \mid t\in[0,T]\} \leq \sum_{i\leq j<\infty} \eta_j(T) \to 0 \quad \text{for } i \to \infty.$$

It is straightforward to show that the function u is a mild solution of the initial-boundary value problem (1)(2)(3) for initial data u_0 on the time interval $[0,T]$. Hence (i) is proved. To show (ii), note that the interval $[0,T]$ was determined at the beginning of the proof uniformly for all initial data $u_0 \in L_\infty(\Omega, R^m)$ such that $\|u_0\|_\infty \leq U_0$.

(iii) Suppose that (iii) is violated. Then there exists $U_0 \in (0,\infty)$ and a sequence (t_n) such that

$$\lim_{n\to\infty} t_n = T_{max} < \infty \quad \text{and} \quad \sup_{n\in N} \|u(t_n)\|_\infty \leq U_0.$$

Hence there exists by (ii) a number $\tau\in(0,\infty)$ and mild solutions v_n,

$$v_n: (x,t)\in\Omega\times[t_n, t_n+\tau) \to v_n(x,t)\in R^m \quad \text{for all } n \in N,$$

for initial data $u(t_n)$ on the intervals $[t_n, t_n+\tau)$. (See Definition 2(d) p. 49 for the meaning). Hence by uniqueness we get a mild solution u for initial data u_0 on the larger interval $[0, T_{max}+\tau)$. This is a contradiction. Hence (iii) must hold.

(iv) For $\iota = 1\ldots m$ define the functions v_ι, G_ι

$v_\iota: (x,t)\in\bar{\Omega}\times[0,T_{max}) \to v_\iota(x,t)\in R$ and $G_\iota: (x,t)\in\bar{\Omega}\times[0,T_{max}) \to G_\iota(x,t)\in R$

by setting

$$v_\iota(x,t) = (P_\iota(t)u_{0\iota})(x) \tag{18}$$

$$G_\iota(x,t) = F_\iota(x,t,u(x,t)). \tag{19}$$

With these definitions the integral equation (6) for the mild solution can be written (omitting again ".,": $v(s) = v(.,s)$, $G(s) = G(.,s)$):

$$u_\iota(t) = v_\iota(t) + \int_0^t P_\iota(t-s)G_\iota(s)ds \quad \text{for } \iota = 1\ldots m, \text{ all } t\in(0,T_{max}). \tag{20}$$

We argue separately for diffusing and nondiffusing components.

First consider the diffusing components. Thus let $\iota = r+1\ldots m$.
Let $\nu,\mu\in(0,2)$ with $\nu/2+\mu < 1$ be given.
We choose $\beta\in(0,1)$, $p\in(1,\infty)$ such that $\nu/2 + n/p < \beta$ and $\mu + \beta < 1$. (21)

The solution of the ι-th equation $u_{\iota t} - d_\iota \Delta u_\iota + d_\iota u_\iota = 0$ of the decoupled linear system with boundary conditions (2b) or (2c) defines the formal semigroup $P_\iota(t)$ in $L_\infty(\Omega)$. Remembering Lemma 1 p.15 we extend this semigroup to an analytic semigroup in the space $L_p(\Omega)$. Denote this analytic semigroup by $e^{-A_{p\iota}t}$ and its generator by $A_{p\iota}$.
For all $\gamma \in (0,1)$ semigroup theory (see e.g. Henry [24], Theorem 1.4.3, p.26) implies the estimates (omitting ι)

$$\|A_p^{-(1-\gamma)}[e^{-A_p t} - I]\|_{p,p} \le K_1 m(t)^{1-\gamma} \quad \text{for all } t \in (0,\infty), \tag{22}$$

$$\|A_p^{\gamma} e^{-A_p t}\|_{p,p} \le K_2 m(t)^{-\gamma} \quad \text{for all } t \in (0,\infty). \tag{23}$$

Let $T \in (0,T_{max})$ be arbitrary. We have $0 < t < t+h \le T < T_{max}$ in the estimates below.
To begin with, we show that the initial data are approximated for $t \to 0$. Let $\gamma \in (0,1)$ be arbitrary. Estimate at first the linear part v_ι:

$$\begin{aligned} \|A_{p\iota}^{\gamma}[v_\iota(t) - u_{0\iota}]\|_p &\le \|A_{p\iota}^{-(1-\gamma)}[e^{-A_{p\iota}t} - I]\|_{p,p} \|A_{p\iota} u_{0\iota}\|_p \\ &\le K_3 m(t)^{1-\gamma} \|u_{0\iota}\|_{C^2(\overline{\Omega})} \quad \text{for all } t \in (0,T_{max}). \end{aligned} \tag{24}$$

Using (20) we now estimate the solution u_ι:

$$\begin{aligned} \|A_{p\iota}^{\gamma}[u_\iota(t) - u_{0\iota}]\|_p &\le \|A_{p\iota}^{\gamma}[v_\iota(t) - u_{0\iota}]\|_p + \\ &\quad + \int_0^t \|A_{p\iota}^{\gamma} e^{-A_{p\iota}(t-s)}\|_{p,p} \|G_\iota(s)\|_p \, ds \\ &\le K_4 m(t)^{1-\gamma} [\|u_{0\iota}\|_{C^2(\overline{\Omega})} + \sup_{0\le s\le T} \|G_\iota(s)\|_\infty] \\ &\qquad \text{for all } t \in (0,T]. \end{aligned} \tag{25}$$

At first, we use (25) with $\gamma = \beta$. Then $\mu < 1-\gamma$ by (21). Furtheron (21) and (40) p.27 show that the embedding $D(A_{p\iota}^{\beta}) \subset C^{\nu}(\overline{\Omega})$ is continuous. Hence

$$\|u_\iota(t) - u_{0\iota}\|_{C^\nu(\overline{\Omega})} \le K_5 m(t)^{\mu} [\|u_{0\iota}\|_{C^2(\overline{\Omega})} + \sup_{0\le s\le T} \|G_\iota(s)\|_\infty] \quad \text{for all } t \in (0,T]. \tag{26}$$

Now apply (25) with $\gamma = \mu+\beta < 1$. We get

$$\|A_{p\iota}^{\mu+\beta}[u_\iota(t) - u_{0\iota}]\|_p \le K_6 \, [\|u_{0\iota}\|_{C^2(\overline{\Omega})} + \sup_{0\le s\le T} \|G_\iota(s)\|_\infty] \quad \text{and hence}$$

$$\|A_{p\iota}^{\mu+\beta}u_\iota(t)\|_p \le K_7\ [\|u_{o\iota}\|_{C^2(\overline{\Omega})} + \sup_{0\le s\le T} \|G_\iota(s)\|_\infty] \quad \text{for all } t\in(0,T]. \tag{27}$$

Next we prove that the solution u is continuous with respect to time t. By the integral equation

$$u_\iota(t+h) = e^{-A_{p\iota}h}u_\iota(t) + \int_0^h e^{-A_{p\iota}(h-s)}G_\iota(t+s)\,ds \tag{28}$$

we get the following estimates corresponding to (24) and (25).
At first for the linear part:

$$\begin{aligned}\|A_{p\iota}^{\beta}[e^{-A_{p\iota}h}u_\iota(t) - u_\iota(t)]\|_p &\le \|A_{p\iota}^{-\mu}[e^{-A_{p\iota}h} - I]\|_{p,p}\|A_{p\iota}^{\beta+\mu}u_\iota(t)\|_p \\ &\le K_8 m(h)^{\mu}\|A_{p\iota}^{\beta+\mu}u_\iota(t)\|_p \qquad (29)\\ &\quad \text{for all } t\in(0,T_{max}),\ h\in(0,T_{max}-t).\end{aligned}$$

Now estimate the solution u:

$$\begin{aligned}&\|A_{p\iota}^{\beta}[u_\iota(t+h) - u_\iota(t)]\|_p \\ &\le \|A_{p\iota}^{\beta}[e^{-A_{p\iota}h}u_\iota(t) - u_\iota(t)]\|_p + \int_0^h \|A_{p\iota}^{\beta}e^{-A_{p\iota}(h-s)}\|_{p,p}\|G_\iota(t+s)\|_p\,ds \\ &\le K_9 m(h)^{\mu}\|A_{p\iota}^{\beta+\mu}u_\iota(t)\|_p + K_{10}\left[\int_0^h m(h-s)^{-\beta}\,ds\right]\sup_{0\le s\le T}\|G_\iota(s)\|_\infty \qquad (30)\\ &\le K_{11}m(h)^{\mu}[\|A_{p\iota}^{\beta+\mu}u_\iota(t)\|_p + \sup_{0\le s\le T}\|G_\iota(s)\|_\infty] \quad \text{for all } t\in(0,T],\ h\in(0,T-t].\end{aligned}$$

By (21) and (40) p.27, the embedding $D(A_{p\iota}^{\beta}) \subset C^{\nu}(\overline{\Omega})$ is continuous. Hence (27)(30) imply

$$\|u_\iota(t+h) - u_\iota(t)\|_{C^{\nu}(\overline{\Omega})} \le K_{12}m(h)^{\mu}\ [\|u_{o\iota}\|_{C^2(\overline{\Omega})} + \sup_{0\le s\le T}\|G_\iota(s)\|_\infty] \tag{31}$$
$$\text{for all } t\in(0,T],\ h\in(0,T-t].$$

After defining $u_\iota(.,0) = u_{o\iota}$, estimates (27) and (31) imply that

$$u_\iota \in C^{\mu}([0,T],C^{\nu}(\overline{\Omega})) \quad \text{for all } T\in(0,T_{max}),\ \nu,\mu\in(0,2) \text{ with } \nu/2+\mu < 1.$$

Putting $\nu = \alpha$, $\mu = \alpha/2$ for $\alpha\in(0,1)$ arbitrary, we get finally

$$u_\iota \in C^{\alpha,\alpha/2}(\overline{\Omega}\times[0,T]) \quad \text{for all } T\in(0,T_{max}),\ \iota = r+1\ldots m,\ \alpha\in(0,1). \tag{32}$$

Now we consider the nondiffusing components for $\iota = 1...r$.
Let $w = (u_1,...,u_r)$ be the vector of nondiffusing components.
Considering the diffusing components as known functions of (x,t), we define the function $H: (x,t,w)\in\bar{\Omega}\times[0,T_{max})\times R^r \to H(x,t,w)\in R^r$ by setting

$$H(x,t,w) = (F_1,...,F_r)(x,t,(w,u_{r+1}(x,t),...,u_m(x,t))). \tag{33}$$

The components $\iota = 1...r$ of the integral equation (6) yield

$$w(x,t) = w_0(x) + \int_0^t H(x,s,w(x,s))\,ds \quad \text{for all } (x,t)\in\bar{\Omega}\times[0,T_{max}) \tag{34}$$

Let $T\in(0,T_{max})$ be arbitrary. Assumptions (SF1)(SF2) and (32) imply that the function H is Hölder-Lipschitz continuous:

For all $T\in(0,T_{max})$, $W \in R$, there exists a constant $L(T,W)$ such that

$$|H(x,t,v)-H(y,s,w)| \leq L(T,W)[|x-y|^\alpha + |t-s|^{\alpha/2} + |v-w|] \tag{35}$$
$$\text{for all } (x,t,v),(y,s,w) \in \bar{\Omega}\times[0,T]\times[-W,W]^r.$$

By the construction of the mild solution in (i), there exists a constant W such that

$$|w(x,t)| \leq W \quad \text{for all } (x,t) \in \bar{\Omega}\times[0,T].$$

Hence (34)(35) imply the estimate

$$|w(x,t)-w(y,t)| \leq |w_0(x)-w_0(y)| + TL(T,W)|x-y|^\alpha + \quad \text{for all } x,y \in \bar{\Omega},$$
$$+ L(T,W)\int_0^t |w(x,s)-w(y,s)|\,ds \quad t\in[0,T].$$

Now Gronwall's Lemma implies

$$|w(x,t)-w(y,t)| \leq [|w_0(x)-w_0(y)| + TL(T,W)|x-y|^\alpha]\, e^{TL(T,W)}$$
$$\text{for all } t\in[0,T],\ x,y \in \bar{\Omega}.$$

Hence $w(.,t) \in C^\alpha(\bar{\Omega})$. Now (34)(35) imply $w_t(.,t) \in C^\alpha(\bar{\Omega})$ for all $t\in[0,T]$. Thus we get finally

$$u_\iota \in C^{\alpha,1+\alpha/2}(\bar{\Omega}\times[0,T]) \quad \text{for all } T\in(0,T_{max}),\ \iota = 1...r \tag{36}$$

and $\alpha\in(0,1)$ from assumptions (SF1)(SF2). Hence assertion (9) is proved.

To prove (10), we have to use the classical linear Schauder theory.

In the nonlinearity $F = F(x,t,u)$, we consider $u = u(x,t)$ as a known function. Thus we get the function $G: (x,t) \in \bar{\Omega} \times [0,T_{max}) \to G(x,t) \in R^m$,

$$G(x,t) = F(x,t,u(x,t)).$$

By assumptions (SF1)(SF2) and (32)(36) we get

$$G \in C^{\alpha,\alpha/2}(\bar{\Omega} \times [0,T], R^m) \quad \text{for all } T \in (0,T_{max}). \tag{37}$$

Now we consider the linear initial-boundary value problem for $\iota = r+1 \ldots m$:

$$v_{\iota t} - d_\iota \Delta v_\iota + d_\iota v_\iota = G_\iota(x,t) \quad \text{for all } (x,t) \in \Omega \times (0,T]; \tag{38}$$

$$v_\iota(x,t) = 0 \quad \text{for } \iota = r+1 \ldots s, (x,t) \in \partial\Omega \times (0,T]; \tag{39b}$$

$$b_\iota(x) v_\iota(x,t) + \partial v_\iota / \partial n(x,t) = 0 \quad \text{for } \iota = s+1 \ldots m, (x,t) \in \partial\Omega \times (0,T]; \tag{39c}$$

$$v_\iota(x,0) = u_{0\iota}(x) \quad \text{for all } x \in \Omega. \tag{40}$$

By assumption (S1) we get for the components $\iota = r+1 \ldots s$ with Dirichlet boundary conditions the compatibility conditions of first order, whereas for the components $\iota = s+1 \ldots m$ with third type boundary conditions the compatibility conditions of zero order hold. Hence the classical Schauder estimates (Ladyzenskaja [32], Theorem 5.2,5.3, p.320) imply

$$v_\iota \in C^{2+\alpha,1+\alpha/2}(\bar{\Omega} \times [0,T]) \quad \text{for } \iota = r+1 \ldots m, \text{ all } T \in (0,T_{max}) \text{ and}$$
$$\|v_\iota\|_{C^{2+\alpha,1+\alpha/2}} \leq K_{12}(T)[\|G_\iota\|_{C^{\alpha,\alpha/2}} + \|u_{0\iota}\|_{C^\alpha}]. \tag{41}$$

Here the constant $K_{12}(T)$ depends only on the structure of the linear system (38)(39) and on the exponent α and the time T.

Clearly the function u is a solution of the system (38)(39) in the mild sense. Hence uniqueness of mild solutions implies $v = u$. Thus the function u satisfies (41) and is a classical solution of the problem (38)(39)(40) in the classical sense on the closed domain $\bar{\Omega} \times [0,T]$.

Thus the Theorem is proved.

In the following we outline the procedure by which a priori bounds are used in the examples of part II to prove global existence of solutions of the initial-boundary value problem for irregular initial data. This procedure was already used in Theorem 6 and 7 of part I.

Initial data which are natural for the system considered and for which in a heurisitic way some primary a priori estimates (as e.g. energy- or entropy estimates) can be derived will be called convenient. Denote the (large) set of convenient initial data by $\mathcal{K}$. We assume that $\mathcal{K} = \bigcup\{\mathcal{K}_b \mid b \in R^r\}$ where b denotes the parameters occuring in the definition of $\mathcal{K}$.

The primary a priori estimate can be derived rigorously only if existence of a classical solution u is known from the beginning, hence only for initial data $u_o \in \mathcal{K}_b \cap \mathcal{R}$. Here $\mathcal{R}$ denotes the set of regular initial data given by Definition 1 p.110. The essential step is to apply the machinery of part I in order to improve the primary a priori estimate. Typically, this process leads to the following result:

For all initial data $u_o \in \mathcal{K}_b \cap \mathcal{R}$, any solution u of the initial-boundary value problem on any time interval [0,T] (which is not specified from the beginning!) the following estimates hold:

$$\sup\{\|u(.,t)\|_\infty \mid t\in[T_1,T_2]\} \leq M(b,T_1,T_2) \quad \text{for all } [T_1,T_2] \subset (0,\infty) \tag{A1}$$

$$\int_0^T \|F(.,s,u(.,s))\|_1 \, ds \leq M(b,T) \quad \text{for all } T \in (0,\infty) \tag{A2}$$

$$\|u(.,t) - P(t)u_o\|_p \leq M(b)o(t) \quad \text{and} \quad \lim_{t\to 0} o(t) = 0 \tag{A3}$$

for some $p\in[1,\infty]$.

Here the constants M depend on the system considered including the boundary conditions and the quantities in the brackets. The important point is that M depends on the initial data only via the parameters b, but not on their smoothness, and that M does not depend on the interval [0,T].

More general initial data in $\mathcal{K} \setminus \mathcal{R}$ are approximated by smooth initial data in $\mathcal{R}$. To this end, the following density assumption, which is satisfied in the examples below, is sufficient:

For all initial data $u_o \in \mathcal{K}_b$, there exists $\tilde{b}\in R^r$ and a sequence (u_{on}) in $\mathcal{K}_{\tilde{b}} \cap \mathcal{R}$ such that $\lim_{n\to\infty} \|u_{on} - u_o\|_1 = 0$. (D)

This density assumption (D) corresponds to Definition 3 p.56.

The final step in the construction of global solutions for irregular initial data is summarized in the following Theorem:

<u>Theorem 2</u> (Construction of global solutions for irregular initial data)

Consider the initial-boundary value problem (1)(2)(3). Assume that all components diffuse, i.e. $r = 0$, $d_\iota > 0$ for $\iota = 1...m$. Assume that (B0)(B1) hold for the boundary condition and (SF1)(SF2) for the nonlinearity F.

Furtheron, assume that the estimates (A1)(A2)(A3) and the density assumption (D) explained above are satisfied (To show this, needs the main work in the examples below).

Then for all convenient initial data $u_o \in \mathcal{K}$, the initial-boundary value problem (1)(2)(3) has a global solution on the interval $[0,\infty)$. This solution satisfies

$$\sup\{\|u(.,t)\|_\infty \mid t\in[T_1,T_2]\} < \infty \quad \text{for all } [T_1,T_2] \subset (0,\infty) \qquad (\overline{A1})$$

$$\int_0^T \|F(.,s,u(.,s))\|_1 ds < \infty \quad \text{for all } T \in (0,\infty) \qquad (\overline{A2})$$

$$\lim_{t\to 0} \|u(.,t) - P(t)u_o\|_p = 0 \quad \text{for p as in (A3)} \qquad (\overline{A3})$$

More precisely, the function u is a L_1-mild solution in the sense of Definition 1 p.29, but not necessarily a $E_{\infty,0,\infty}$-mild solution (p.111):

$$u(.,t) = P(t)u_o + \int_0^t P(t-s)F(.,s,u(.,s))\, ds \quad \text{for all } t\in(0,\infty), \qquad (M)$$

where the integral is an absolutely converging Bochner integral in the space $L_1(\Omega)$.

The function u is a classical solution, too. More precisely:

$$u \in C^{2+\alpha,1+\alpha/2}(\overline{\Omega}\times[T_1,T_2],R^m) \quad \text{for all intervals } [T_1,T_2] \subset (0,\infty). \quad (S)$$

For all $(x,t) \in \overline{\Omega}\times(0,\infty)$, (including $x \in \partial\Omega$, but excluding $t = 0$!) the function u satisfies the differential equation (1).

For all $(x,t) \in \partial\Omega\times(0,\infty)$, (excluding $t = 0$) the function u satisfies the boundary conditions (2).

Remark

Assumption (A2) may be dropped. Then all conclusions of the Theorem remain valid besides $(\overline{A2})$. The integral in (M) has to be interpretated as an improper integral in the sense

$$\int_0^t = L_\infty\text{-}\lim_{\tau\to 0} \int_\tau^t .$$

Proof

For initial data $u_0 \in \mathcal{K}$, there exists $b \in R^r$ and an approximating sequence (u_{on}) in $\mathcal{K}_b \cap \mathcal{R}$ such that

$$\lim_{n\to\infty} \| u_{on} - u_o \|_1 = 0. \tag{D}$$

By Theorem 1(iv) there exist classical solutions u_n for the regular initial data u_{on} on some time intervals $[0,T_n)$. Let $T_n = T_{max}$ be the maximal existence time. Then the explosion property (8) and the uniform a priori estimate (A1) imply that $T_n = \infty$ for all $n \in N$. Hence there exist global solutions for the whole sequence u_{on} of approximative initial data. Indeed, (A1) implies the uniform estimate (we drop .,)

$$\sup \{ \| u_n(t) \|_\infty \mid t \in [T_1,T_2],\ n \in N \} \le M(T_1,T_2) \quad \text{for all } [T_1,T_2] \subset (0,\infty)$$

and for the nonlinear term F

$$\sup \{ \| F(t,u_n(t)) \|_\infty \mid t \in [T_1,T_2], n \in N \} \le M(T_1,T_2) \quad \text{for all } [T_1,T_2] \subset (0,\infty),$$

too.We apply Lemma 21 p.78 to the equations (1) for all components $\iota = 1...m$. Since all components diffuse, we get by (262)

$$\sup_{n \in N} \| u_n \|_{C^\mu([T_1,T_2],C^\nu(\overline{\Omega},R^m))} < \infty \quad \text{for all } [T_1,T_2] \subset (0,\infty), \text{ all } \nu,\mu \text{ with } \nu/2 + \mu < 1 \tag{37}$$

Now we argue as in the proof of Theorem 6 p.94. By the Theorem of Ascoli the embedding $C^{\tilde\mu}([T_1,T_2],C^{\tilde\nu}(\overline{\Omega})) \subset C^\mu([T_1,T_2],C^\nu(\overline{\Omega}))$ is compact if $\tilde\mu > \mu$, $\tilde\nu > \nu$. Hence applying a diagonal argument with respect to T_1,T_2, ν,μ the uniform estimate (37) shows that there exists a subsequence (u_{n_j}) of (u_n) and a function $u: (x,t) \in \overline{\Omega} \times (0,\infty) \to u(x,t) \in R^m$ such that

$$\lim_{n\to\infty} \| u_{n_j} - u \|_{C^\mu([T_1,T_2],C^\nu(\overline{\Omega},R^m))} = 0 \quad \text{for all } [T_1,T_2] \subset (0,\infty), \ \nu,\mu \text{ with } \nu/2 + \mu < 1. \tag{38}$$

For all $t \in (0,\infty)$, the sequences $u_j(t)$, $F(t,u_j(t))$ and by Lemma 3 p.25 and (D) the sequence $P(t)u_{oj}$, too, converge in the norm of $C^\nu(\overline{\Omega},R^m)$.

(We simply write u_j for the subsequence u_{n_j}).Hence the estimates (A1) (A2)(A3) for the approximating solutions u_j imply $(\overline{A1})(\overline{A2})(\overline{A3})$ for the solution u.

To show (M), note that the approximating solutions u_j are $E_{\infty,o,\infty}$-mild solutions in the sense of Definition 2 p.111.
Splitting the integral in (6) at some (small) $\tau\in(0,t)$ yields

$$u_j(t) = P(t)u_{oj} + \int_\tau^t P(t-s)F(s,u_j(s))ds + R_j(\tau) \tag{39}$$

with the "rest"

$$R_j(\tau) = P(t-\tau)[P(\tau)u_{oj} - u_j(\tau)]. \tag{40}$$

The limit for $j \to \infty$ exists of all terms in the norm of $C^\nu(\overline{\Omega},R^m)$. We get

$$u(t) = P(t)u_o + \int_\tau^t P(t-s)F(s,u(s))ds + R(\tau) \tag{41}$$

with the "rest"

$$R(\tau) = P(t-\tau)[P(\tau)u_o - u(\tau)]. \tag{42}$$

In the limit $\tau \to 0$, the rest can be estimated

$$\|R(\tau)\|_\infty \le \|P(t-\tau)\|_{\infty,p}\|P(\tau)u_o - u(\tau)\|_p. \tag{43}$$

Hence Lemma 3 p.25 and estimate $(\overline{A3})$ imply

$$\lim_{\tau\to 0} \|R(\tau)\|_\infty = 0. \tag{44}$$

Thus we get from (41)(44)

$$u(t) = P(t)u_o + L_\infty\text{-}\lim_{\tau\to 0} \int_\tau^t P(t-s)F(s,u(s))ds. \tag{45}$$

By estimate (A2) the integral may also be interpretated as an absolutely converging Bochner integral in $L_1(\Omega)$. Thus (M) is proved.

It remains to show (S) i.e. that we get classical solutions for $t > 0$. By Theorem 1(iv), the approximating solutions u_j are smooth classical solutions which satisfy the differential equation (1) and the boundary condition (2) on the closed domains $\overline{\Omega}\times[0,\infty)$ and $\partial\Omega\times[0,\infty)$, respectively. We proceed as in the proof of Theorem 1 p.118. In the nonlinearity F, we consider $u_j = u_j(x,t)$ as a known function. Thus define the functions

$$G_j(x,t) = F(x,t,u_j(x,t)) \quad\text{and}\quad G(x,t) = F(x,t,u(x,t)).$$

Now (38) and (SF1)(SF2) imply

$$\lim_{j\to\infty} \|G_j - G\|_{C^{\alpha,\alpha/2}(\bar{\Omega}\times[T_1,T_2],R^m)} = 0 \quad \text{for all } [T_1,T_2] \subset (0,\infty).$$

We take $T_1 > 0$ as initial time. The classical Schauder estimates for the linear problem (38)(39) imply (Ladyzenskaja [32], p.320):

$$\|u_i - u_j\|_{C^{2+\alpha,1+\alpha/2}(\bar{\Omega}\times[T_1,T_2],R^m)} \leq K(T_1,T_2)\|G_i - G_j\|_{C^{\alpha,\alpha/2}(\bar{\Omega}\times[T_1,T_2],R^m)}$$
$$\text{for all } i,j \in N, \text{ all } [T_1,T_2] \subset (0,\infty).$$

Hence the sequence (u_j) converges in the Banach space of differentiable functions occuring on the left-hand side. Since the approximating functions u_j are classical solutions, the limit function u is a classical solution, too. It is essential for this argument to assume $T_1 > 0$.

Thus the Theorem is proved.

<u>Theorem 3</u> (Comparison of solutions by the strong maximum principle)

Let the domain $\Omega \subset R^N$ be bounded with boundary $\partial\Omega \in C^2$. Let the functions $v, \rho_k \in C^1(\partial\Omega)$ $k = 1...N$ satisfy

$$b(x) \geq 0 \quad \text{for all } x \in \partial\Omega \tag{46}$$

$$\inf \{ \sum_{1\leq k\leq N} \rho_k(x) n_k(x) \mid x \in \partial\Omega\} > 0 , \tag{47}$$

where $n_k(x)$ $k = 1...N$ denotes the outer normal unit vector at the boundary point $x \in \partial\Omega$.

Let the function $F: (x,t,u) \in \bar{\Omega}\times[0,T]\times R \to F(x,t,u) \in R$ be continuous together with the partial derivative $\partial F/\partial u$.

If the functions $u,v \in C^{2,1}(\bar{\Omega}\times[0,T])$ satisfy the following inequalities (48)[(49a)or(49b)](50):

$$u(x,0) \leq v(x,0) \quad \text{for all } x \in \Omega \tag{48}$$

$$v(x,t) \leq v(x,t) \quad \text{for all } x \in \partial\Omega,\ t \in (0,T) \tag{49a}$$

$$b(x)u(x,t)+\partial u/\partial\rho(x,t) \leq b(x)v(x,t)+\partial v/\partial\rho(x,t) \tag{49b}$$
$$\text{for all } x \in \partial\Omega,\ t \in (0,T)$$

$$u_t - \Delta u - F(x,t,u) \leq v_t - \Delta v - F(x,t,v) \quad \text{for all } x \in \Omega,\ t \in (0,T); \tag{50}$$

then either $u \equiv v$ or (51a) - in case of boundary conditions (49b) even (51b) - and (52) hold.

$$u(x,t) < v(x,t) \quad \text{for all } x \in \overline{\Omega} \setminus \partial\Omega,\ t \in (0,T] \tag{51a}$$

$$u(x,t) < v(x,t) \quad \text{for all } x \in \overline{\Omega},\ t \in (0,T] \tag{51b}$$

$$\begin{array}{ll} \text{If} & u(y,t) = v(y,t) \quad \text{for some } y \in \partial\Omega, t \in (0,T], \\ \text{then} & \partial u/\partial\rho(y,t) > \partial v/\partial\rho(y,t). \end{array} \tag{52}$$

Proof

Since u and v are given functions on the domain $(x,t) \in \overline{\Omega} \times [0,T]$, there exists $L \in (0,\infty)$ and a continuous function

$$C: (x,t) \in \overline{\Omega} \times [0,T] \to C(x,t) \in [0,\infty) \quad \text{such that for all } (x,t) \in \overline{\Omega} \times [0,T]$$

$$F(x,t,u(x,t)) - F(x,t,v(x,t)) = (L - C(x,t))(u(x,t) - v(x,t)).$$

The function $w: (x,t) \in \overline{\Omega} \times [0,T] \to w(x,t) \in R$ defined by

$$w(x,t) = e^{-Lt}(u(x,t) - v(x,t))$$

satisfies (53)[(54a)or(54b)](55):

$$w \leq 0 \quad \text{for all } x \in \Omega\ ,\ t = 0\ ; \tag{53}$$

$$w \leq 0 \quad \text{for all } x \in \partial\Omega,\ t \in (0,T) \tag{54a}$$

or

$$bw + \partial w/\partial\rho \leq 0 \quad \text{for all } x \in \partial\Omega,\ t \in (0,T); \tag{54b}$$

$$w_t - \Delta w + Cw \leq 0 \quad \text{for all } x \in \Omega\ ,\ t \in (0,T). \tag{55}$$

Define $M := \max \{w(x,t) \mid (x,t) \in \overline{\Omega} \times [0,T]\}$.

We apply Protter and Weinberger [49], p.173, section 3 to the inequality (55). Hence either (56) or (57)(58) hold:

$$w(x,t) \equiv M \quad \text{for all } x \in \overline{\Omega},\ t \in [0,T] \tag{56}$$

or

$$w(x,t) < M \quad \text{for all } x \in \overline{\Omega} \setminus \partial\Omega,\ t \in (0,T] \tag{57}$$

and

$$\begin{array}{l} w(y,t) = M \quad \text{for some } y \in \partial\Omega,\ t \in (0,T] \quad \text{implies} \\ \partial w/\partial\rho(y,t) > 0. \end{array} \tag{58}$$

Especially, the function w attains the maximum M on the parabolic boundary:

$$M = \max \{w(x,t) \mid (x,t) \in \overline{\Omega}\times\{0\}\cup\partial\Omega\times[0,T]\}. \qquad (59)$$

Next we show that $M \leq 0$.

Indeed, for Dirichlet boundary conditions (a), we know by (53)(54a) that the function w is nonpositive on the parabolic boundary. Consider the case of third type boundary conditions (b). If the maximum M is attained for $t = 0$, (53) implies $M \leq 0$. Assume that the maximum M is attained for some $(y,t) \in \partial\Omega\times(0,T]$ and (56) does not hold. (58) **implies**

$$w(y,t) = M \quad \text{and} \quad \partial w/\partial\rho(y,t) > 0.$$

Hence (54b) implies

$$b(y)M < b(y)w(y,t) + \partial w/\partial\rho(y,t) \leq 0 \qquad (60)$$

Since $b(y) \geq 0$ by assumption (46), this implies $M \leq 0$.

Hence we have shown $M \leq 0$.

If $M < 0$ all assertions of the Theorem are trivially satisfied. Hence we may assume $M = 0$. From the alternative - either (56) or (57)(58) - we conclude that either $u \equiv v$ or (51a)(52) hold. It remains to show (51b) in the case (b) of third type boundary conditions. To argue by contradiction, assume that $u(y,t) = v(y,t)$ for some $(y,t)\in\partial\Omega\times(0,T]$. Then $w(y,t) = 0$ and (60) yields the **contradiction** $0 < 0$.

Thus the Theorem is proved.

<u>We recall some conventions</u>:

If $u = u(x,t)$ is some function $(x,t)\in\Omega\times(0,T) \to u(x,t)\in R$, we denote by $u(t)$ or $u(.,t)$ the function $x \in \Omega \to u(x,t)\in R$. Whether ".," is dropped or not is more or less chance. We hope that no confusion arises.

The space dimension is N. Let $n = N/2$. The denominator after the fraction bar / consists always of a single number, unless brackets occur.

Constants denoted by the letter K depend only on the exponents and the domain Ω. Other quantities on which K depends are indicated in brackets.

The letter M denotes constants depending on the exponents, the domain Ω and some further quantities specified in each proof below.

Let $m(t) = \min\{1,t\}$ for all $t\in[0,\infty)$.

Let $\displaystyle ⨍_{\Omega} = \frac{1}{|\Omega|} \int_{\Omega}$.

The Gierer-Meinhardt Model

In the study of various topics from developmental biology, Gierer and Meinhardt [38,39] proposed the following system of reaction-diffusion equations:

$$u_t - D\Delta u + \mu u = u^2/v + \rho \qquad \text{for all } x \in \Omega,\ t > 0. \tag{1a}$$

$$v_t - \overline{D}\Delta v + \nu v = u^2 + \overline{\rho} \tag{1b}$$

Here u and v are the concentrations of two substances, which are usually called activator and inhibitor. The peaks of high concentration of v give the positional information for the development of structures, e.g. an insect embryo.

Returning to mathematical terms, $\Omega \subset R^N$ is a bounded domain. Let $n = N/2$ for simpler notation. Furtheron, let u and **v** be functions $u: (x,t)\in\overline{\Omega}\times[0,T] \to u(x,t)\in R$ and $v: (x,t)\in\overline{\Omega}\times[0,T] \to v(x,t)\in R$.

We assume Neumann boundary conditions

$$\frac{\partial u}{\partial n} = 0 \quad \text{and} \quad \frac{\partial v}{\partial n} = 0 \qquad \text{for all } x \in \partial\Omega,\ t > 0 \tag{1c}$$

and initial conditions

$$u(x,0) = u_o(x), \quad v(x,0) = v_o(x) \qquad \text{for all } x \in \Omega. \tag{1d}$$

The diffusion rates $D,\overline{D} \in (0,\infty)$ usually satisfy $D \ll \overline{D}$. Furtheron, $\nu,\mu,\rho,\overline{\rho} \in [0,\infty)$ are constants. The mathematically more simple case is $\nu,\mu,\rho+\overline{\rho} \in (0,\infty)$. The limiting cases violating this need some special considerations.

Theorem (Globally bounded solutions for space dimension N = 1,2,3)

Take space dimension N = 1,2 or 3 and let $p_o,q_o\in[1,\infty]$ satisfy $(n = N/2)$

$$1/p_o < 1/2 + 1/(2n), \quad 1/p_o + 1/q_o < 1/n, \quad 1/q_o < 1/n - 1/2. \tag{2}$$

Assume that the initial data $u_o \in L_{p_o}(\Omega)$ and $v_o \in L_1(\Omega)$ satisfy

$$u_o \geq 0,\ v_o \geq 0,\ \|1/v_o\|_{q_o} < \infty. \tag{3}$$

Furtheron assume $D,\bar{D}\in(0,\infty)$, $\nu,\mu,\rho,\bar{\rho}\in[0,\infty)$ and (to exclude degenerate cases)

$$\|u_o\|_{p_o} + \rho > 0 \tag{4}$$

$$\text{If} \quad \rho = \bar{\rho} = 0, \text{ then } \nu\mu = 0. \tag{5}$$

Then the Gierer-Meinhardt system (1a)(1b)(1c) has a global classical solution on the time interval $(0,\infty)$ satisfying (instead of (1d))

$$\lim_{t\to 0} \|u(t) - u_o\|_{p_o} = 0 \quad \text{if } p_o < \infty \text{ or } p_o = \infty \text{ and } u_o \in C(\bar{\Omega}); \tag{6a}$$

$$\lim_{t\to 0} \|v(t) - v_o\|_1 = 0. \tag{6b}$$

Furthermore, if $\nu,\mu > 0$, then

$$\sup_{0<t<\infty} m(t)^{\delta}\|u(t)\|_p < \infty \quad \text{for all } p\in[p_o,\infty],\ \delta = n/p_o - n/p; \tag{7a}$$

$$\sup_{0<t<\infty} m(t)^{\eta}\|v(t)\|_q < \infty \quad \text{for all } q\in[1,\infty],\ \eta = n - n/q. \tag{7b}$$

If $\nu\mu = 0$, the estimates analogous to (7a)(7b) hold on all bounded intervals $[0,T]$ with arbitrary $T\in(0,\infty)$.

<u>Proof</u>

The set of convenient initial data is $\mathcal{K} = \cup\{\mathcal{K}_b \mid b\in(0,\infty)\}$ where $\mathcal{K}_b$ is the set of all $(u_o,v_o) \in L_{p_o} \times L_1$ such that (2)(4) hold and

$$u_o \geqq 0, \quad v_o \geqq 0 \quad \text{and} \quad \|u_o\|_{p_o} + \|v_o\|_1 + \|1/v_o\|_{q_o} \leqq b. \tag{8}$$

In the followings Lemmas we can restrict ourselves to regular convenient initial data in $\mathcal{K} \cap \mathcal{R}$ (see Definition 1 p.110). By Theorem 1 p.111, the system (1a)(1b)(1c)(1d) has a classical solution (u,v) on $[0,T_{max})$.

Lemma 1 gives a simple initial-boundary value problem (9a)-(9d) for lower bounds $\underline{u},\underline{v}$ of the solution u,v.

Lemma 2 gives explicit estimates for the lower bound $\underline{v}$.

Lemma 3 constructs the primary a priori estimate for the component u. For initial data $u_o \in L_2$, this turns out to be an L_2-estimate. We sketch the idea leading to this estimate. Consider the system

$$u_t = u^2/v \quad \text{and} \quad v_t = u^2,$$

$$u(0) = u_o \quad \text{and} \quad v(0) = v_o.$$

This ordinary initial-value problem with initial data $u_o \geq 0$, $v_o > 0$ is the backbone of the Gierer-Meinhardt system. One gets immediately

$$\frac{d}{dt}(u - \log v) = 0 \quad \text{and hence} \quad u(t) = \log v(t) + C \quad \text{for all } t > 0.$$

Hence the differential equation for v implies

$$u(t) \leq \log\left[v_o + \int_0^t u^2(\tau)d\tau\right] + \overline{C} \quad \text{for all } t > 0.$$

Since the left-hand side grows less for $u \to \infty$ than the right-hand side, there exists $C_1 \in (0,\infty)$ such that

$$u(t) \leq C_1 \quad \text{for all } t > 0.$$

In Lemma 3, we exploit this idea for the whole Gierer-Meinhardt system.

Lemma 4 uses Theorem 6 of part I to derive L_∞-estimates for the component u from the primary L_2-estimate of Lemma 3.

Lemma 5 derives the L_∞-estimates for the component v.

Lemma 1

Let $\underline{u} = \underline{u}(x,t)$, $\underline{v} = \underline{v}(x,t)$ be the solution of the initial-boundary value problem

$$\underline{u}_t - D\Delta\underline{u} + \mu\underline{u} = \rho \qquad (9a)$$

for all $x \in \Omega$, $t > 0$

$$\underline{v}_t - \overline{D}\Delta\underline{v} + \nu\underline{v} = \underline{u}^2 + \overline{\rho} \qquad (9b)$$

with Neumann boundary conditions

$$\partial\underline{u}/\partial n = 0 \quad \text{and} \quad \partial\underline{v}/\partial n = 0 \quad \text{for all } x \in \partial\Omega,\ t > 0 \qquad (9c)$$

and initial conditions

$$\underline{u}(x,0) = u_o(x), \quad \underline{v}(x,0) = v_o(x) \quad \text{for all } x \in \Omega. \qquad (9d)$$

Furtheron, assume that

$$\|u_o\|_{p_o} + \rho > 0. \qquad (4)$$

Then the solution (u,v) of the system (1) can be estimated from below:

$$u(x,t) > \underline{u}(x,t) > 0 \quad \text{and} \quad v(x,t) > \underline{v}(x,t) > 0 \qquad (10)$$

for all $x \in \overline{\Omega}$, $t \in (0,T_{max})$.

Proof For the equation (9a) and the functions $\underline{u}$ and 0, the Comparison Theorem p.123 yields

$$\underline{u}(x,t) > 0 \quad \text{for all } x \in \overline{\Omega},\ t > 0, \text{ in shorthand } \underline{u} \supset 0.$$

Application of the Comparison Theorem to equation (9b) and the functions $\underline{v}$ and 0 yields

$$\underline{v}(x,t) > 0 \quad \text{for all } x \in \overline{\Omega},\ t > 0.$$

Application to the equation (1a) and the functions u and $\underline{u}$ yields

$$u(x,t) \geqq \underline{u}(x,t) \quad \text{for all } x \in \overline{\Omega},\ t \geqq 0.$$

Application to the equation (1b) and the functions v and $\underline{v}$ yields

$$v(x,t) \geqq \underline{v}(x,t) \quad \text{for all } x \in \overline{\Omega},\ t \geqq 0.$$

Hence we get $u \supset 0$ and $v \supset 0$. Repetition of the argument finally yields the strict inequality $u \supset \underline{u}$ and $v \supset \underline{v}$.

Thus the Lemma is proved.

Lemma 2

Let $q_0, q_1, q_2 \in [1,\infty]$ satisfy

$$n/q_0 < n/q_1 + 1/q_2. \tag{11}$$

Then there exists a constant K (depending only on the exponents q_0, q_1, q_2 and on $D, \overline{D}$ and on Ω) such that the solution $(\underline{u},\underline{v})$ of (9) satisfies

$$\|1/\underline{v}(t)\|_{q_0} \leqq K\, e^{\nu t} \|1/v_0\|_{q_0} \quad \text{for all } t > 0, \tag{12}$$

$$\|1/\underline{v}\|_{q_1,q_2,T} \leqq K\, e^{\nu T} \|1/v_0\|_{q_0} \quad \text{for all } T > 0, \tag{13}$$

$$\min\{\underline{v}(x,t) \mid x \in \overline{\Omega}\} \geqq K^{-1} m(t)^n e^{-\nu t} \|1/v_0\|_1^{-1} \quad \text{for all } t > 0, \tag{14}$$

$$\min\{\underline{v}(x,t) \mid x \in \overline{\Omega}\} \geqq \begin{cases} \rho^2/[8(\nu+\mu)^2] + \overline{\rho}/[2(\nu+\mu)] & \text{if } e^{-(\nu+\mu)t} \leqq 1/2 \\ \overline{\rho} t + \rho^3 t^3/3 & \text{if } (\nu,\mu) = (0,0). \end{cases} \tag{15}$$

Assume that $\rho+\overline{\rho} > 0$ or $\nu = 0,\ v_0 \neq 0$ or $\mu = 0,\ u_0 \neq 0$. Then

$$\min\{\underline{v}(x,t) \mid x \in \overline{\Omega},\ t \geqq 1\} \geqq b_2 > 0. \tag{16}$$

Proof

The function $w = 1/\underline{v}(x,t)$ satisfies the differential equation

$$\left[\frac{\partial}{\partial t} - \overline{D}\Delta - \nu\right] w = -(\underline{u}^2 + \overline{\rho}) w^2 - 2\overline{D}(\nabla w)^2/w \leqq 0 \tag{17}$$

and Neumann boundary conditions $\partial w/\partial n = 0$ on $\partial\Omega$. Denote by $S(t)$ the semigroup generated by the operator $-\overline{D}\Delta$ in Ω with Neumann boundary conditions on $\partial\Omega$. Since the right-hand side of the differential equation (17) is nonpositive, the Comparison Theorem p.123 yields

$$0 \leqq w(x,t) \leqq e^{\nu t}[S(t)(1/v_o)](x) \quad \text{for all } x \in \overline{\Omega},\ t \geqq 0.$$

Now apply estimate (30) of Lemma 3 p.25.
Hence we get for all $q_o, q_1 \in [1,\infty]$,, $\delta = \max\{0, n/q_o - n/q_1\}$ that

$$\| 1/\underline{v}(t) \|_{q_1} \leqq K_{13} m(t)^{-\delta} e^{\nu t} \| 1/v_o \|_{q_o} \quad \text{for all } t > 0.$$

For $q_o = q_1$ we get (12), whereas for $q_o = 1$, $q_1 = \infty$ we get (14).
If q_o, q_1, q_2 satisfy assumption (11), we have $\delta < 1/q_2$. Hence (13) follows by a simple integration.
Let the functions $u_1 = u_1(t)$, $v_1 = v_1(t)$ be the solution of the ordinary initial value problem

$$u_{1t} + \mu u_1 = \rho \quad \text{and} \quad v_{1t} + \nu v_1 = u_1^2 + \overline{\rho}\ ,$$

$$u_1(0) = 0 \quad \text{and} \quad v_1(0) = 0.$$

The Comparison Theorem p.123 implies

$$\underline{u}(x,t) \geqq u_1(t) \quad \text{and} \quad \underline{v}(x,t) \geqq v_1(t) \quad \text{for all } x \in \overline{\Omega},\ t \geqq 0.$$

Computing u_1 and v_1 explicitly yields the estimate (15).

It remains to show (16).
If $\rho+\overline{\rho} > 0$, this is immediate from (15). Hence assume $\rho = \overline{\rho} = 0$.
If $\nu = 0$, we have $v_o \geqq 0$, $v_o \not\equiv 0$ by assumption and $\underline{v}_t - \overline{D}\Delta\underline{v} \geqq 0$ by (9b). Hence the asymptotic behavior of the linear Neumann problem implies (16).
If $\mu = 0$, we have $u_o \geqq 0$, $u_o \not\equiv 0$ by assumption and $\underline{u}_t - D\Delta\underline{u} \geqq 0$ by (9a). By the behavior of the linear Neumann problem we get

$$\min\{\underline{u}(x,t) \mid x \in \overline{\Omega},\ t \geqq 1/2\} > 0.$$

Application of the Comparison Theorem to (9b) yields (16).

Thus the Lemma is proved.

<u>Lemma 3</u>
Let $p_o, r_2 \in [1,\infty]$ and $\delta \in (0,1/2)$ satisfy

$$n/p_o - n/2 \leqq \delta \quad \text{and} \quad n/p_o - n/2 < 1/r_2. \tag{18}$$

Take convenient regular initial data $u_o \in L_{p_o}$, $v_o \in L_{q_o}$ satisfying

$$u_o \geqq 0\,,\quad v_o \geqq 0\,,\quad \|u_o\|_{p_o} + \|v_o\|_1 + \|1/v_o\|_1 \leqq b_1 \tag{19}$$

and via the solution $(\underline{u},\underline{v})$ of problem (9)

$$\min\{\underline{v}(x,t) \mid x \in \overline{\Omega},\ t \geqq 1\} \geqq b_2 > 0. \tag{20}$$

(By (16) of Lemma 2, this holds e.g. under the assumptions (4)(5):

$$\|u_o\|_{p_o} + \rho > 0 \quad \text{and} \quad \text{if } \rho = \overline{\rho} = 0, \text{ then } \nu\mu = 0.\ ) \qquad (4)(5)$$

Then there exists a constant M - depending on the exponents p_o, r_2, δ, the domain Ω, the constant $D, \overline{D}, \nu, \mu, \rho, \overline{\rho}$ in the equations (1a)(1b), the constants b_1, b_2 for above and (only **in exceptional** cases $\nu = 0$ or $\mu = 0$) on the time T - such that the solution (u,v) of the Gierer-Meinhardt system (1a)(1b)(1c)(1d) on any time interval [O,T] satisfies the following a priori estimates:

$$\|u\|_{2,\delta,T} \leqq M(T) \tag{21}$$

$$|||u|||_{2,r_2,T} \leqq M(T) \tag{22}$$

If $\nu\mu \neq 0$, the estimates (21)(22) even hold with M independent of T.

<u>Proof</u>

In the differential equations (1a)(1b) we introduce new variables

$U = U(x,t)$ by $U = u(x,t) - \underline{u}(x,t)$ and $V = V(x,t)$ by $V = \log v(x,t)$.

Lemma 1 implies that

$U(x,t) \geqq 0$ and $V(x,t)$ is well defined for all $x \in \overline{\Omega}$, $t \in [0, T_{max})$.

The Gierer-Meinhardt system (1a)(1b) is transformed into

$$U_t - D\Delta U + \mu U = u^2/v \tag{23}$$

$$V_t - \overline{D}\Delta V = (u^2 + \overline{\rho})/v - \nu + \overline{D}(\nabla V)^2 \tag{24}$$

for all $x \in \overline{\Omega}$, $t \in [0, T_{max})$.

We get homogenous Neumann boundary conditions on $\partial\Omega$ for U and V. The function U - V satisfies the differential equation

$$\left[\frac{\partial}{\partial t} - D\Delta + \mu\right](U - V) = (D - \overline{D})\Delta V - \overline{D}(\nabla V)^2 + \nu - \overline{\rho}/v - \mu V \tag{25}$$

$$\leqq (D - \overline{D})\Delta V + \nu - \mu V$$

for all $x \in \overline{\Omega}$, $t \in [0, T_{max})$.

and homogenous Neumann boundary conditions on $\partial\Omega$.

Let the function $a: (x,t) \in \bar{\Omega} \times [0,T_{max}) \to a(x,t) \in R$ be the solution of the initial-boundary value problem

$$\left[\frac{\partial}{\partial t} - D\Delta + \mu\right] a = (D-\bar{D})\Delta V \quad \text{for all } x \in \bar{\Omega},\ t \in (0,T_{max}); \tag{26a}$$

$$\frac{\partial a}{\partial n} = 0 \quad \text{for all } x \in \partial\Omega,\ t \in (0,T_{max}); \tag{26b}$$

$$a(x,0) = 0 \quad \text{for all } x \in \bar{\Omega}. \tag{26c}$$

Then (25)(26) imply

$$\left[\frac{\partial}{\partial t} - D\Delta + \mu\right](U - V - a) \leq \nu - \mu V \quad \text{for all } x \in \bar{\Omega},\ t \in [0,T_{max}) \tag{27a}$$

and homogenous Neumann boundary conditions

$$\frac{\partial}{\partial n}(U - V - a) = 0 \quad \text{for all } x \in \partial\Omega,\ t \in (0,T_{max}). \tag{27b}$$

Denote by $S(t)$ the semigroup generated by $-D\Delta$ in Ω with Neumann boundary conditions on $\partial\Omega$. Let $V_o \in C(\bar{\Omega})$ be given by $V_o = V_o(x) = \log v_o(x)$ for all $x \in \bar{\Omega}$. By integration of (27) we get

$$\begin{aligned} 0 \leq U(x,t) &\leq a(x,t) + \nu(1-e^{-\mu t})/\mu + e^{-\mu t}[S(t)V_o](x) \\ &\quad + V(x,t) - \mu\left[\int_0^t S(t-s)e^{-\mu(t-s)}V(.,s)\,ds\right](x) \\ &\qquad \text{for all } x \in \bar{\Omega},\ t \in [0,T_{max}). \end{aligned} \tag{28}$$

Next we estimate the function a. The differential equation (26a) and boundary condition (26b) imply (recall $⨍_\Omega = 1/|\Omega| \int_\Omega$)

$$\begin{aligned} \left[\frac{d}{dt} + 2\mu\right] ⨍_\Omega a^2(x,t)\,dx + 2D ⨍_\Omega (\nabla a)^2 dx &= 2(D-\bar{D}) ⨍_\Omega (\Delta V) a\,dx \\ &= -2(D-\bar{D}) ⨍_\Omega \nabla V\, \nabla a\,dx \leq 2|D-\bar{D}|\,\|\nabla V\|_2 \|\nabla a\|_2 \\ &\leq (D-\bar{D})^2 (2D)^{-1} \|\nabla V(.,t)\|_2^2 + 2D\|\nabla a(.,t)\|_2^2 \quad \text{for all } t \in [0,T_{max}). \end{aligned} \tag{29}$$

A useful estimate of $\|\nabla V(t)\|_2^2$ can be derived by integrating the differential equation (24) and using the Neumann boundary condition on $\partial\Omega$:

$$\frac{d}{dt} ⨍_\Omega V(x,t)\,dx \geq -\nu + \bar{D}\|\nabla V(.,t)\|_2^2 \quad \text{for all } t \in [0,T_{max}). \tag{30}$$

Define $d = (D-\bar{D})^2/(2D\bar{D})$. Subtracting (30) from (29) implies

$$\left[\frac{d}{dt} + 2\mu\right]\int_\Omega (a^2 - dV)(x,t)\,dx \le d\nu - 2d\mu\int_\Omega V(x,t)\,dx \quad \text{for all } t\in[0,T_{max})$$

and hence by integration with respect to time

$$\|a(.,t)\|_2^2 \le d\|V_o\|_1 + d\nu(1-e^{-2\mu t})/(2\mu) + d\int_\Omega V(x,t)\,dx + \\ + 2d\mu\int_0^t e^{-2\mu(t-s)}\left(\int_\Omega V(x,s)\,dx\right) ds \quad \text{for all } t\in[0,T_{max}). \tag{31}$$

From (28) and (31) we get estimates of $\|U(.,t)\|_2$ and finally $\|U\|_{2,\delta,T}$:

$$\|U(.,t)\|_2 \le$$

$$\le \|a(.,t)\|_2 + \|V_o\|_2 + \|V(.,t)\|_2 + \nu(1-e^{-\mu t})/\mu + \int_0^t e^{-\mu(t-s)}\|V(.,s)\|_2 ds$$

$$\le 1 + (1+d)\left[\|V_o\|_2 + \|V(.,t)\|_2 + \nu(1-e^{-\mu t})/\mu + 2\mu\int_0^t e^{-\mu(t-s)}\|V(.,s)\|_2 ds\right] \quad \text{for all } t\in[0,T_{max})$$

and

$$\|U\|_{2,\delta,T} \le 1 + (1+D^{-1}+\bar{D}^{-1})\left[\|V_o\|_2+\nu\min\{1/\mu,T\}+2(1+\mu)(1-\delta)^{-1}\|V\|_{2,\delta,T}\right] \quad \text{for all } T\in(0,T_{max}). \tag{32}$$

Next we estimate $\|V\|_{2,\delta,T}$ occuring on the right-hand side.
Define the functions

$$h: t\in[0,T_{max}) \to h(t)\in[1,\infty) \quad \text{and} \quad H: (x,t)\in\bar{\Omega}\times[0,T_{max}) \to H(x,t)\in R$$

by

$$h(t) = \max\{1, e/\underline{v}(x,t) \mid x \in \bar{\Omega}\} \quad \text{and} \quad H(x,t) = h(t)v(x,t).$$

Note that by Lemma 1, we have $v \ge \underline{v} > 0$ and hence $H \ge e$.
Since the function $\varphi = \varphi(H) = \log^2 H$ satisfies

$$\varphi'(H) \ge 0 \quad \text{and} \quad \varphi''(H) \le 0 \quad \text{for all } H\in[e,\infty),$$

Jensen's inequality

$$\int_\Omega \varphi(H)\,dx \le \varphi\left(\int_\Omega H\,dx\right)$$

implies that the function $V = \log v = \log(H/h)$ satisfies

$$\int_\Omega V^2 dx = \int_\Omega (\log^2 H + \log^2 h - 2\log H \log h)\,dx \le \log^2\left(\int_\Omega H dx\right) + \log^2 h.$$

Taking roots we get

$$\|V(.,t)\|_2 \le \log\int_\Omega H dx + \log h \le \log\int_\Omega v(x,t)\,dx + 2\log h(t) \quad \text{for all } t\in[0,T_{max}).$$

For simpler notation, define $\log_+ y := \max\{0, \log y\}$.
Multiplying by $m(t)^\delta$, the last estimate implies

$$\|V\|_{2,\delta,T} \leq \log_+\|v\|_{1,0,T} + 2\sup\{m(t)^\delta \log h(t) \mid t\in[0,T]\} \quad (33)$$
$$\text{for all } T\in(0,T_{max}).$$

Estimates (14) of Lemma 2 and assumption (20) imply

$$\log h(t) \leq K + \nu t + n\log(1/m(t)) + \log_+\|1/v_o\|_1 \quad \text{for all } t > 0, \quad (34)$$

$$\log h(t) \leq 1 + \log_+(1/b_2) \quad \text{for all } t \geq 1. \quad (35)$$

Estimates (33)(34)(35) together imply

$$\|V\|_{2,\delta,T} \leq \log_+\|v\|_{1,0,T} + K + \nu + n/(e\delta) + \log_+(1/b_2) + \log_+\|1/v_o\|_1$$
$$\text{for all } T\in(0,T_{max}).(36)$$

To estimate $\|v\|_{1,0,T}$, note that the differential equation (1b) and the Neumann boundary condition (1c) for the function v imply

$$\left[\frac{d}{dt} + \nu\right]\int_\Omega v(x,t)dx = \bar{\rho} + \|u(.,t)\|_2^2 \quad \text{for all } t\in(0,T_{max})$$

and hence by integration with respect to time t

$$\int_\Omega v(x,t)dx \leq \|v_o\|_1 + \bar{\rho}(1-e^{-\nu t})/\nu + \int_0^t e^{-\nu(t-s)}\|u(.,s)\|_2^2\, ds$$
$$\text{for all } T\in(0,T_{max}).$$

Denote by S(t) the semigroup generated by $-D\Delta$ in Ω with Neumann boundary conditions on $\partial\Omega$. Equation (9a) implies

$$\underline{u}(.,t) = e^{-\mu t}S(t)u_o + \rho(1-e^{-\mu t})/\mu = u(.,t) - U(.,t) \quad \text{for all } t > 0.$$

Hence estimate (30) of Lemma 3(i) p.25 and (18) imply for all t > 0

$$\|u(.,t)\|_2 \leq m(t)^{-\delta}(K_{13}e^{-\mu t}\|u_o\|_{p_o} + \|U\|_{2,\delta,t}) + \rho(1-e^{-\mu t})/\mu. \quad (36a)$$

Hence

$$\|v\|_{1,0,T} \leq \|v_o\|_1 + K_1[\min\{1/\nu,T\} + 2\delta/(1-2\delta)](\|u_o\|_{p_o}^2 + \|U\|_{2,\delta,T}^2) +$$
$$+ \bar{\rho}\min\{1/\nu,T\} + 4\rho^2\int_0^T e^{-\nu(T-s)}\min\{1/\mu^2, s^2\}ds \quad (37)$$
$$\text{for all } T\in(0,T_{max}).$$

Here the constants K depend only on the exponents p_o, δ and on D and Ω.

For the rest of the proof, we denote by the letter M constants which depend on the domain Ω and on the following arguments in the indicated domain:

$p_o \in [1,\infty]$; $\delta \in (0,1/2) \cap [n/p_o - n/2, 1/2)$; $D, \bar{D} \in (0,\infty)$; $\rho, \bar{\rho} \in [0,\infty)$; $\min\{1/\nu, T\}$, $\min\{1/\mu, T\} \in [0,\infty)$; $b_2 \in (0,\infty)$.

Note that M can be chosen independent of time T except in the cases $\nu = 0$ or $\mu = 0$.

Now we can write the important estimates (32)(36)(37) in a simplified form. To eliminate $V_o = \log v_o$ occuring in (32), note that

$$\log^2 v \leq 2(v + 1/v) \quad \text{for all } v \in (0,\infty) \quad \text{and hence}$$

$$\|V_o\|_2 \leq 1 + \|v_o\|_1 + \|1/v_o\|_1.$$

Hence we get

$$\|U\|_{2,\delta,T} \leq M(1 + \|v_o\|_1 + \|1/v_o\|_1 + \|V\|_{2,\delta,T}) \tag{38}$$

$$\|V\|_{2,\delta,T} \leq M(1 + \|1/v_o\|_1 + \log_+ \|v\|_{1,0,T}) \tag{39}$$

$$\|v\|_{1,0,T} \leq M(1 + \|u_o\|_{p_o}^2 + \|v_o\|_1 + \|U\|_{2,\delta,T}^2) \tag{40}$$

for all $T \in (0, T_{max})$.

which imply

$$1 + \|U\|_{2,\delta,T} \leq M_1(1 + \|u_o\|_{p_o} + \|v_o\|_1 + \|1/v_o\|_1 + \log(1 + \|U\|_{2,\delta,T})) \quad \text{for all } T \in (0, T_{max}).$$

We may assume $M_1 > 1$. With b_1 given by (20) and $y = 1 + \|U\|_{2,\delta,T}$ we get

$$y \leq M_1(1 + b_1 + \log y). \tag{41}$$

Let the function $x: b \in [0,\infty) \to x(b) \in (0,\infty)$ be defined as the solution of

$$x(b) = M_1(1 + b + \log x(b)) \qquad \text{for all } b \in [0,\infty).$$

Implicit differentiation yields

$$x'(b) = M_1 x(b)/(x(b) - M_1) \leq M_1 + M_1/b \quad \text{for all } b \in (0,\infty), \tag{42}$$

since $\quad x(b) > M_1(1 + b) \qquad$ for all $b \in [0,\infty)$.

On the other hand, one can check that

$$x(b) \leq M_1\, e^{2M_1} \quad \text{for all } b \in [0,1]. \tag{43}$$

Finally we get by (42)(43)

$$y \leq x(b) \leq M_1(e^{2M_1} + 2b) \qquad \text{for all } b \in [0,\infty)$$

proving

$$\|U\|_{2,\delta,T} \leq M_2(1 + \|u_o\|_{p_o} + \|v_o\|_1 + \|1/v_o\|_1) \quad \text{for all } T \in (0, T_{max}). \tag{44}$$

Hence (36a) implies

$$\|u\|_{2,\delta,T} \leq M_3(1 + \|u_o\|_{p_o} + \|v_o\|_1 + \|1/v_o\|_1) \quad \text{for all } T\in(0,T_{max}) \tag{45}$$

which proves assertion (21).

To show (22), let $\delta < 1/r_2$. Then the simple estimate

$$|\!|\!| u |\!|\!|_{2,r_2,T} \leq \left[\int_0^1 m(s)^{-r_2\delta}\, ds\right]^{1/r_2} \|u\|_{2,\delta,T}$$

yields the assertion (22). Thus the Lemma is proved.

<u>Lemma 4</u>

Make the assumptions of the Theorem p.126 and let the initial data u_o and v_o be regular.

Then there exists $\varepsilon\in(0,1)$ and a constant M depending on the exponents $p_o, q_o, p, \delta, \varepsilon$, the domain Ω, the quantities $D, \overline{D}, \nu, \mu, \rho, \overline{\rho}$, time T and

$$\|u_o\|_{p_o} + \|v_o\|_1 + \|1/v_o\|_{q_o} = b \quad \text{and} \quad \min\{\underline{v}(x,t) \mid x \in \overline{\Omega}, t \geq 1\} = b_2$$

such that the following estimates hold for the solution u,v of the Gierer-Meinhardt system (1a)(1b)(1c)(1d):

$$\|u(.,t) - S(t)u_o\|_{p_o} \leq m(t)^{\varepsilon} M(T) \qquad \text{for all } t\in(0,T], \tag{47}$$

$$\sup_{0<t\leq T} m(t)^{\delta}\|u(.,t)\|_p \leq M(T) \qquad \text{for } p\in[p_o,\infty], \text{ all } T\in(0,T_{max}) \text{ and } \delta = n/p_o - n/p. \tag{48}$$

(Here S(t) denotes the semigroup generated by $-D\Delta$ in Ω with Neumann boundary conditions on $\partial\Omega$).

The constant M can be chosen independent of T except for the limiting case $\nu\mu = 0$.

<u>Proof</u>

There exist $q_1, q_2, r_2 \in [1,\infty]$ such that

$$n/q_o < n/q_1 + 1/q_2, \tag{49}$$

$$n/p_o - n/2 < \min\{1/2, 1/r_2\}, \tag{50}$$

$$n/q_1 + 1/q_2 + n/r_1 + 1/r_2 < 1, \tag{51}$$

$$1/q_1 + 1/r_1 < 1. \tag{52}$$

We denote by M_i constants as specified in the Lemma.

Take the first equation (1a) of the Gierer-Meinhardt system and consider the function $c = 1/v(x,t)$ appearing in the right-hand side $u^2/v + \rho$ as a known weight function.

By the short time estimate (13) and the long time estimate (16) (which we can apply by the assumptions (4)(5)) Lemma 2 yields

$$|||c|||_{q_1,q_2,T} \leq K e^{\nu} \|1/v_o\|_{q_o} + 1/b_2 \leq M_1 \quad \text{for all } T \in (0,T_{max}). \tag{53}$$

Note that M_1 does not depend on T due to the long time estimate (16). By Lemma 3 we have the estimate

$$|||u|||_{r_1,r_2,T} \leq M_2(T) \quad \text{for all } T \in (0,T_{max}). \tag{54}$$

with $r_1 = 2$.

Let $\gamma = 2$. To the equation (1a) we apply Lemma 19 and 20 of part I. By (51)(52) there exist $\beta, \varepsilon \in (0,1)$ such that the assumptions (231)(232) (233) of Lemma 20 are satisfied. Hence (234) implies

$$\|u(.,t)\|_{\infty} \leq m(t)^{-n/p_o} M_3(T) \qquad \text{for all } t \in (0,T] \subset (0,T_{max}). \tag{55}$$

Furtheron there exist p, δ such that the assumptions (207) - (211) of Lemma 19 are satisfied (this follows by Lemma 16). Hence (214)(215) imply

$$\|u(.,t) - S(t)u_o\|_{p_o} \leq m(t)^{\varepsilon} M_4(T) \quad \text{for all } t \in (0,T] \subset (0,T_{max}). \tag{56}$$

Now (55)(56) prove (47)(48) via the interpolation (124). Unless the exceptional case $\nu\mu = 0$ occurs, M_2 and hence M_3, M_4 can be chosen independent of T. Thus the Lemma is proved.

Lemma 5

Make the assumptions of the Theorem p.126 and let the initial data u_o and v_o be regular.

Then there exists $\varepsilon \in (0,1)$ and a constant M as specified in Lemma 4 such that the following estimates hold for the solution u,v of the Gierer-Meinhardt system (1a)(1b)(1c)(1d):

$$\|v(.,t) - v_o\|_1 \leq m(t)^{\varepsilon} M_5(T) \qquad \text{for all } t \in (0,T], \tag{57}$$

$$\sup_{0<t\leq T} m(t)^{\eta} \|v(.,t)\|_q \leq M_5(T) \text{ for all } q \in [1,\infty], \text{ all } T \in (0,T_{max}) \tag{58}$$
$$\eta = n - n/q.$$

The constant M can be chosen independent of time T except in the case that $\nu\mu = 0$.

Proof

There exists a finite sequence $1 = q_0 < q_1 < \dots < q_k = \infty$ such that

$$n/q_{i-1} - n/q_i = \delta_i < 1 \quad \text{for } i = 1\dots k, \tag{59}$$

$$\max\{0, 2n/p_0 - n/q_i\} = \overline{\delta}_i \quad \text{for } i = 0\dots k \text{ and} \tag{60}$$

$$\overline{\delta}_0 < \min\{1, \delta_2, \delta_3 \dots \delta_k\}. \tag{61}$$

The second equation (1b) of the Gierer-Meinhardt system is

$$\left[\frac{\partial}{\partial t} - \overline{D}\Delta + \nu\right] v = f, \tag{62}$$

where the inhomogenous term $f = f(x,t) = \overline{\rho} + u^2(x,t)$ satisfies by (48)

$$\|f(.,t)\|_{q_i} \leq M(T) m(t)^{-\overline{\delta}_i} \quad \text{for all } t \in (0,T] \subset (0,T_{max}),\ i = 0\dots k. \tag{63}$$

Denote by $S(t)$ the semigroup generated by $-\overline{D}\Delta+\nu$ with Neumann boundary conditions on $\partial\Omega$. The mild formulation of (62)

$$v(.,t) = S(t)v_0 + \int_0^t S(t-s) f(.,s) ds \quad \text{for all } t \in (0,T_{max}) \tag{64}$$

implies the estimates

$$\begin{aligned} \|v(.,t)\|_{q_1} &\leq \|S(t)\|_{q_1,1} \|v_0\|_1 + \int_0^t \|S(t-s)\|_{q_1,1} \|f(.,s)\|_1 \, ds, \\ \|v(.,t) - S(t)v_0\|_1 &\leq \int_0^t \|S(t-s)\|_{1,1} \|f(.,s)\|_1 \, ds \quad \text{for all } t \in (0,T_{max}). \end{aligned} \tag{65}$$

We use estimate (30) of Lemma 3(i). Hence (63) with $i = 0$ and (65) imply imply the assertion (57) with $\varepsilon = 1 - \overline{\delta}_0 > 0$.

We prove by induction on i that

$$\|v\|_{q_i, n-n/q_i, T} \leq M_i(T) \quad \text{for all } T \in (0,T_{max}),\ i = 1\dots k. \tag{66}$$

For $i = 1$, this follows by (65).

Assume that (66) is already proved for some $i-1 \in \{1\dots k-1\}$. Taking t_1 as initial time and $v(.,t_1)$ as initial data, we get by (62)

$$v(.,t_1+t_2) = S(t_2)v(.,t_1) + \int_0^{t_2} S(t_2-s) f(.,t_1+s) \, ds \tag{67}$$

$$\text{for all } t_1 \in [0,T_{max}),\ t_2 \in (0, T_{max}-t_1).$$

Hence by (30) of Lemma 3(i) part I, (63) and (66) for i-1, we estimate

$$\|v(.,2t)\|_{q_i} =$$

$$\leq \|S(t)\|_{q_i,q_{i-1}} \|v(.,t)\|_{q_{i-1}} + \int_0^t \|S(t-s)\|_{q_i,q_{i-1}} \|f(.,t+s)\|_{q_{i-1}} ds$$

$$\leq K_{13} m(t)^{-(n/q_{i-1}-n/q_i)} e^{-\nu t} M_{i-1}(T) m(t)^{-n(1-1/q_{i-1})} + \tag{68}$$

$$+ K_{13} M(T) \int_0^t e^{-\nu(t-s)} m(t-s)^{-\delta_i} m(t+s)^{-\overline{\delta}_{i-1}} ds \quad \text{for all } 2t \in (0,T].$$

Since by assumption (2) we have $1/p_o \leq 1/2 + 1/2n$, it is easy to check that $\delta_i + \overline{\delta}_{i-1} - 1 \leq n(1 - 1/q_i)$. Hence evaluation of the integral in (68) shows the assertion (66) with i (even for all $T < T_{max}$!).

Finally, (66) for i = k yields the assertion (58).

In the more common case $\nu > 0$, the constants M in (63)(66) can be chosen independent of T because of the factor $e^{-\nu t}$ in the integrals.

Thus the Lemma is proved.

We finish the proof of the Theorem according to the process outlined p.119 - 123. Take arbitrary convenient initial data u_o, v_o. Then there exists a sequence u_{om}, v_{om} of regular convenient initial data and constants $b, b_2 \in (0,\infty)$ such that

$$\lim_{m\to\infty} [\|u_{om}-u_o\|_1 + \|v_{om}-v_o\|_1] = 0,$$

$$\sup_{m\in N} \|u_{om}\|_{p_o} + \|v_{om}\|_1 + \|1/v_{om}\|_{q_o} \leq b,$$

$$\inf \{\underline{v}_m(x,t) \mid x \in \overline{\Omega},\ t \geq 1,\ m \in N\} \geq b_2 > 0,$$

where $\underline{v}_m$ are the lower solutions given by (9) for initial data u_{om}, v_{om}. The estimates (47)(48) of Lemma 4 and (57)(58) of Lemma 5 hold uniformly in $m \in N$. Hence the assertions of the Theorem follow by the arguments of Theorem 2 p.120.

Thus the proof of the Theorem is finished.

The Brusselator

In the study of chemical systems far from equilibrium, the following simple reaction-diffusion system was proposed by Prigogine and Glansdorff [48]:

$$u_t - a\Delta u = A - (B+1)u + u^2 v \qquad \text{for all } x \in \Omega,\ t > 0. \tag{1a}$$

$$v_t - b\Delta v = Bu - u^2 v \tag{1b}$$

Here $a,b,B\in(0,\infty)$ and $B\in[0,\infty)$ are constants. In the following section, we consider the system (1) in a bounded domain $\Omega \subset R^N$ with Neumann boundary conditions

$$\frac{\partial u}{\partial n} = 0 \quad \text{and} \quad \frac{\partial v}{\partial n} = 0 \qquad \text{for all } x \in \partial\Omega,\ t > 0 \tag{1c}$$

and initial conditions

$$u(x,0) = u_o(x) \quad \text{and} \quad v(x,0) = v_o(x) \quad \text{for all } x \in \Omega. \tag{1d}$$

We restrict ourselves to initial data $u_o \geqq 0$, $v_o \geqq 0$.

Theorem 1 (Globally bounded solutions for space dimension $N = 1,2,3$)

Take space dimension $N = 1,2$ or 3 and let $p_o,q_o\in[2,\infty]$ satisfy $(n = N/2)$

$$n/q_o < 1 - n/2. \tag{2}$$

Assume for the initial data $u_o \in L_{p_o}(\Omega)$, $v_o \in L_{q_o}(\Omega)$ and $u_o,v_o \geqq 0$.

Then the Brusselator system (1a)(1b)(1c) has a global classical solution on the time interval $(0,\infty)$ satisfying

$$\lim_{t\to 0} \| u(.,t) - u_o\|_{p_o} = 0 \quad \text{if } p_o < \infty \text{ or } p_o = \infty \text{ and } u_o \in C(\bar{\Omega}); \tag{3a}$$

$$\lim_{t\to 0} \| v(.,t) - v_o\|_1 = 0; \tag{3b}$$

$$\sup_{0<t<\infty} m(t)^\delta \| u(.,t)\|_p \leqq M \quad \text{for all } p\in[p_o,\infty],\ \delta = n/p_o - n/p; \tag{4a}$$

$$\sup_{0<t<\infty} m(t)^\eta \| v(.,t)\|_q \leqq M \quad \text{for all } q\in[q_o,\infty],\ \eta = n/q_o - n/q. \tag{4b}$$

Here the constant M depends continuously on the exponents p_o, q_o, on the domain Ω, the constants a,b,A.B in the differential equations (1) and on $\|u_o\|_{p_o}$ and $\|v_o\|_{q_o}$.

Proof

The set of convenient initial data consists of all $(u_o, v_o) \in L_{p_o} \times L_{q_o}$ with $p_o, q_o \in [2,\infty]$ such that (2) holds and $u_o, v_o \geqq 0$.

In the following Lemmas, we can restrict ourselves to regular convenient initial data (see Definition 1 p 110). By Theorem 1 p.111, the system (1a)(1b)(1c)(1d) has a classical solution (u,v) on the maximal interval $[0, T_{max})$.

Lemma 1 gives the primary a priori estimates. These are an L_2-estimate for the component u and an L_∞-estimate for the component v.

Lemma 2 applies the results of part I (Lemma 19,20 and Theorem 6) in order to derive uniform estimates.

Lemma 1

Take convenient regular initial data u_o, v_o.

Then there exists a constant M depending on the domain Ω and on a,b,A,B and the norms $\|u_o\|_2$ and $\|v_o\|_2$ such that the solution of the Brusselator system (1a)(1b)(1c)(1d) satisfies the following estimates:

$$\|u(.,t)\|_2 \leq M \qquad \text{for all } t \in (0, T_{max}) \tag{5}$$

$$\int_0^t \|\nabla u(.,s)\|_2^2 + \|u(.,s)\|_2^2 \, ds \leq M(1+t) \qquad \text{for all } t \in (0, T_{max}) \tag{6}$$

$$\int_0^t \|u + u^2 v(.,s)\|_1 \, ds \leq M(1+t) \qquad \text{for all } t \in (0, T_{max}) \tag{7}$$

$$v^2(x,t) \leq [S(t) v_o^2](x) + M(A,B) \qquad \text{for all } x \in \overline{\Omega},\ t \in (0, T_{max}). \tag{8}$$

Here S(t) denotes the semigroup generated by $-b\Delta$ with Neumann boundary conditions on $\partial\Omega$.

Proof

Let α: $t \in [0,\infty) \to \alpha(t) \in R$ be the solution of the initial value problem

$$\frac{d\alpha}{dt}(t) + (B+1)\alpha(t) = A\,, \qquad \alpha(0) = 0.$$

The Comparison Theorem p.123 and explicit computation of α yield

$$u(x,t) \geq \alpha(t) = A(1-e^{-(B+1)t})/(B+1) \quad \text{for all } x \in \bar{\Omega},\ t\in[0,T_{max}). \tag{9}$$

By (1b) the function $w = v^2(x,t)$ satisfies the differential equation

$$\begin{aligned}\left[\frac{\partial}{\partial t} - b\Delta\right] w &= 2(B\ uv - (uv)^2) - 2b(\nabla v)^2 \quad \text{for all } x \in \Omega, t\in(0,T_{max}) \\ &\leq \varphi(u^2 w) \leq \varphi(\alpha^2(t) w(x,t)).\end{aligned} \tag{10}$$

Here the nonincreasing function $\varphi\colon y\in[0,\infty) \to \varphi(y)\in R$ is defined by

$$\varphi(y) = \max_{z^2 \geq y} 2(Bz - z^2) = \begin{cases} 2(B\sqrt{y} - y) & \text{for } y \geq B^2/4 \\ B^2/2 & \text{for } y \leq B^2/4. \end{cases}$$

Let $\beta\colon t\in[0,\infty) \to \beta(t)\in R$ be the solution of the initial value problem

$$\frac{d\beta}{dt}(t) = \varphi(\alpha^2(t)\beta(t)), \quad \beta(0) = 0.$$

Some calculations show that β satisfies

$$\beta(t) \leq B^2 t/2 \quad \text{for all } t\in[0,\infty),$$

$$\lim_{t\to\infty} \varphi(\alpha^2(t)\beta(t)) = 0, \qquad \lim_{t\to\infty} \beta(t) = (AB/(B+1))^2 \quad \text{and hence}$$

$$\beta(t) \leq M(A,B) \quad \text{for all } t\in[0,\infty). \tag{11}$$

Since the function $\bar{w}\colon (x,t)\in\bar{\Omega}\times[0,T_{max}) \to \bar{w}(x,t)\in R$ given by $\bar{w}(x,t) = [S(t)v_0^2](x) + \beta(t)$ satisfies the differential equation

$$\left[\frac{\partial}{\partial t} - b\Delta\right]\bar{w} = \varphi(\alpha^2(t)\beta(t)) \geq \varphi(\alpha^2(t)\bar{w}(x,t)) \quad \text{for all } x \in \Omega, t\in(0,T_{max}) \tag{12}$$

and both w and $\bar{w}$ satisfy Neumann boundary conditions on $\partial\Omega$, (10)(12) and the Comparison Theorem p.123 imply $w \leq \bar{w}$. Hence more explicitly

$$\begin{aligned} v^2(x,t) &\leq [S(t)v_0^2](x) + \beta(t) \\ &\leq [S(t)v_0^2](x) + M(A,B), \end{aligned} \qquad \text{for all } x \in \bar{\Omega},\ t\in[0,T_{max})$$

where (11) has been used in the last step. Thus assertion (8) is proved.

Next we estimate the component u. Adding the equations (1a)(1b) yields

$$\frac{\partial}{\partial t}(u+v) - \Delta(au+bv) = A - u \quad \text{for all } x \in \Omega,\ t\in(0,T_{max}).$$

Multiplying by $(u+v)$ and integrating over the domain Ω, we get

$$\frac{d}{dt}\int_\Omega (u+v)^2/2\, dx + \int_\Omega [a(\nabla u)^2 + (a+b)(\nabla u)(\nabla v) + b(\nabla v)^2]dx$$
$$= \int_\Omega (A-u)(u+v)\, dx \quad \text{for all } t\in(0,T_{max}). \tag{13}$$

If the diffusion coefficients a and b are not equal, the quadratic form in ∇u and ∇v on the right-hand side is indefinite. Yet we can use that

$$a(\nabla u)^2 + (a+b)(\nabla u)(\nabla v) + b(\nabla v)^2 \geq a(\nabla u)^2/2 - (a^2+b^2)(\nabla v)^2/(2a). \tag{14}$$

To handle the term ∇v, note that the equation (1b) for v implies

$$\frac{d}{dt}\int_\Omega v^2/2\, dx + \int_\Omega b(\nabla v)^2 dx = \int_\Omega uv(B-uv)\, dx \leq B^2/4 \quad \text{for all } t\in(0,T_{max}). \tag{15}$$

Hence (13)(14)(15) imply that the function

$W: (x,t)\in\bar{\Omega}\times[0,T_{max}) \to W(x,t)\in R$ given by $W = (u+v)^2 + (a^2+b^2)v^2/(2ab)$

satisfies the estimate

$$\frac{d}{dt}\int_\Omega W\, dx + a\int_\Omega (\nabla u)^2 dx \leq M(a,b,B) + \int_\Omega 2(A-u)(u+v)dx \quad \text{for all } t\in(0,T_{max}). \tag{16}$$

Since (8) implies by integration over the domain Ω that

$$\|v(.,t)\|_2 \leq \|v_0\|_2 + M(A,B) \quad \text{for all } t\in(0,T_{max}),$$

we get

$$\left[\frac{d}{dt} + 1\right]\int_\Omega W\, dx + \int_\Omega [u^2 + a(\nabla u)^2]dx \leq M_1 + \int_\Omega 2(A+v)(u+v)dx$$
$$\leq M_1 + 2\|A+v\|_2\|u+v\|_2 \leq M_1 + 2\|A+v\|_2^2 + \|u+v\|_2^2/2 \leq M_1 + (1/2)\int_\Omega W\, dx \quad \text{for all } t\in(0,T_{max})$$

and finally

$$\left[\frac{d}{dt} + 1/2\right]\int_\Omega W\, dx + \int_\Omega [u^2 + a(\nabla u)^2]dx \leq M_1 \quad \text{for all } t\in(0,T_{max}). \tag{17}$$

Integration with respect to time t yields

$$\|u(.,t)\|_2^2 \leq \int_\Omega W(x,t)dx \leq 2M_1 + \int_\Omega W(x,0)dx \quad \text{for all } t\in(0,T_{max}). \tag{18}$$

Hence (5) is proved. Integration of (17) with respect to time yields (6), too. Integration of the equation (1b) for the component v with respect to time t and over the domain Ω yields

$$\|v(.,t)\|_1 + \int_0^t\int_\Omega u^2v - Bu\, dx \leq \|v_0\|_1 \quad \text{for all } t\in(0,T_{max}). \tag{19}$$

Hence (6) implies the assertion (7). Thus the Lemma is proved.

Lemma 2

Make the assumptions of the Theorem 1 and let the initial data (u_o, v_o) be regular.

Then there exists $\varepsilon \in (0,1)$ and a constant M_2 depending on the exponents p_o, q_o, the domain Ω, on a,b,A,B and on the norms $\|u_o\|_{p_o}$, $\|v_o\|_{q_o}$ such that the following estimates hold:

$$\|u(.,t) - S_a(t)u_o\|_{p_o} \leq m(t)^{\varepsilon} M_2 \text{ for all } t \in (0,T_{max}), \tag{20}$$

$$\|v(.,t) - S_b(t)v_o\|_1 \leq m(t)^{\varepsilon} M_2 \text{ for all } t \in (0,T_{max}), \tag{21}$$

$$\sup_{0<t<T_{max}} m(t)^{\delta} \|u(.,t)\|_p \leq M_2 \text{ for all } p \in [p_o,\infty],\ \delta = n/p_o - n/p, \tag{22}$$

$$\sup_{0<t<T_{max}} m(t)^{\eta} \|v(.,t)\|_q \leq M_2 \text{ for all } q \in [1,\infty],\ \eta = n/q_o - n/q. \tag{23}$$

Here $S_a(t)$ and $S_b(t)$ denote the semigroups generated by $-a\Delta$ and $-b\Delta$ with Neumann boundary conditions on $\partial\Omega$.

Proof

First we consider equation (1a) for the component u.

By the assumptions of Theorem 1, there exist $q_1, q_2 \in [1,\infty]$ such that

$$n/q_1 + 1/q_2 < 1 - n/2 \quad \text{and} \quad 1/q_1 < 1/2, \tag{24}$$

$$n/q_o < n/q_1 + 1/q_2. \tag{25}$$

Furtheron, let $r_1 = 2$, $r_2 = \infty$ and $\gamma = 2$.

In the equation (1a) for the component u, we consider the second component v occuring on the right-hand side as a known weight function c. By estimate (8) of Lemma 1 and (30) of Lemma 3(i), part I and (25), we get

$$|||c|||_{q_1,q_2,T} \leq K\|v_o\|_{q_o} + M(A,B) \quad \text{for all } T \in (0,T_{max}). \tag{26}$$

On the other hand, estimate (5) of Lemma 1 shows that

$$|||u|||_{r_1,r_2,T} \leq M \qquad\qquad \text{for all } T \in (0,T_{max}). \tag{27}$$

We apply Lemma 19 and 20 of part I. By (24) there exist $\beta, \varepsilon \in (0,1)$ such that the assumptions (231)(232)(233) of Lemma 20 are satisfied. Furtheron, there exist p, δ such that the assumptions (207) - (211) of

Lemma 19 are satisfied (this follows by Lemma 16). Hence assertions (214)(215) of Lemma 19 prove (20). Assertion (234) of Lemma 20 proves (22) for $p = \infty$. The general case $p \in [p_o, \infty]$ follows by the interpolation (124) p.51.

Now consider the component v.
Estimate (8) of Lemma 1 and (30) of Lemma 3(i), part I imply (23).
It remains to show that v approaches the initial data in the sense of (21). In the sharper sense $\|v(.,t)-v_o\|_{q_o} \to 0$ for $t \to 0$, this leads to further restrictions on p_o and hence the initial data. Thus we consider in (21) only the L_1-norm.

Since $2/p_o + 1/q_o < 1/n + 1$, there exists $\varepsilon \in (0,1)$, $p \in [p_o, \infty]$, $q \in [q_o, \infty]$ such that

$$2/p + 1/q \leq 1$$

and

$$0 < 2\delta + \eta + \varepsilon < 1 \quad \text{with} \quad \eta = n/q_o - n/q\,, \quad \delta = n/p_o - n/p. \tag{28}$$

The differential equation (1b) for the component v and (22)(23) imply

$$\begin{aligned} \|v(.,t)-S_b(t)v_o\|_1 &\leq \int_0^t \|S_b(t-s)\|_{1,1} \|(Bu-u^2v)(.,s)\|_1 \, ds \leq \\ &\leq \int_0^t \|(Bu+u^2v)(.,s)\|_1 ds \leq \int_0^t m(t)^{-(2\delta+\eta)} ds (BM_2 + M_2^3) \leq \\ &\leq [m(t)^{\varepsilon} + t]M_3 \quad \text{for all } t \in (0, T_{max}). \end{aligned} \tag{29}$$

Hence (21) follows for $t \in (0,1)$. The rest is clear by (23).
Thus the Lemma is proved.

Now the proof of the Theorem is finished by the process outlined p.119 through 123. Arbitrary convenient initial data (u_o, v_o) can be approximated by a sequence (u_{om}, v_{om}) of regular convenient initial data in the sense of (D) p.119. Then the assertions of the Theorem follow by the arguments of Theorem 2 p.120. Thus the proof is finished.

Using a Lemma of Kielhöfer, which is the essential tool in [30], we prove global existence of solutions even for space dimension N = 4.

<u>Theorem 2</u> (Global solution for space dimension N = 4)

Let $\Omega \subset R^4$ be a bounded domain with smooth boundary. Take nonnegative initial data $u_o \in L_4(\Omega)$ and $v_o \in L_\infty(\Omega)$.

Then the Brusselator system (1a)(1b)(1c)(1d) has a global classical solution (u,v) on the time interval $(0,\infty)$. Furtheron

$$\lim_{t\to 0} \|u(.,t) - u_o\|_4 = 0; \tag{30}$$

$$\lim_{t\to 0} \|v(.,t) - v_o\|_1 = 0; \tag{31}$$

$$\sup_{0<t<T} m(t)^\delta \|u(.,t)\|_p \leq M(T) \quad \text{for all } T\in(0,\infty),\ p\in[4,\infty],\ \delta = 1/2-2/p; \tag{32}$$

$$\sup_{0<t<\infty} \|v(.,t)\|_\infty \leq M. \tag{33}$$

<u>Proof</u>

Estimates (5)(6)(7)(8) hold. Hence (33) is clear. For space dimension N = 4, we have the Sobolev embedding

$$L_4(\Omega) \subset W_2^1(\Omega) \quad \text{and} \quad \|u\|_4 \leq K\|u\|_{W_2^1} \quad \text{for all } u \in W_2^1(\Omega).$$

Hence estimate (6) of Lemma 1 implies

$$\int_0^t \|u(.,s)\|_4^2 \, ds \leq M_4(1 + t) \quad \text{for all } t\in(0,T_{max}), \tag{34}$$

where the constant M_4 depends only on the domain Ω,a,b,A,B and the norm $\|u_o\|_2$ and $\|v_o\|_2$.

By estimate (30) of Lemma 3(i) p.25 we get

$$\|S_a(t)u\|_4 \leq K\, m(t)^{-1/2} e^{-t} \|u\|_2 \quad \text{for all } u \in L_2,\ t\in(0,\infty). \tag{35}$$

Hence the mild formulation of the differential equation (1a)

$$u(.,t) = S_a(t)u_o + \int_0^t S_a(t-s)(A+u^2v)(.,s)ds$$

implies the estimate

$$\|u(.,t)\|_4 \leq K\|u_o\|_4 + M_5\int_0^t m(t-s)^{-1/2} e^{-(t-s)} (1+\|u(.,s)\|_4^2)\, ds \quad \text{for all } t\in(0,T_{max}). \tag{36}$$

By the Lemma of Kielhöfer (appendix of [30]) formulas (34)(36) imply

$$\sup \{\|u(.,t)\|_4 \mid t\in(0,T]\} \leq M_6(T) \quad \text{for all } T\in(0,T_{max}). \tag{37}$$

Here the constant $M_6(t)$ depends on the domain Ω, on a.b.A.B and the norms $\|u_o\|_4$ and $\|v_o\|_\infty$ as well as on the time T.

Let $q_1 = q_2 = \infty$, $r_1 = 4$, $r_2 = \infty$ and $\gamma = 2$. We apply Lemma 19 and 20 to the equation (1a) for the component u and use (37) as a primary a priori estimate. Hence (30)(32) follow. Finally (31) is clear by the argument p.145.
Thus the Theorem is proved analogously to Theorem 1 above.

The FitzHugh-Nagumo System

The following system was proposed by FitzHugh as a model for the spread of a nerve pulse along the nerve axon (see the article of FitzHugh in Schwan [60] for the biological background):

$$v_t = \Delta v + f(v) - u \qquad \text{for all } x \in R^N,\ t > 0. \tag{1a}$$

$$u_t = \sigma v - cu \tag{1b}$$

Here the functions v and u: $(x,t) \in R^N \times [0,\infty) \to v(x,t)$ and $u(x,t) \in R$ can be considered as electric voltage and activation of the membrane. σ and c are positive constants. Usually, one assumes $N = 1$, because the axon can be considered as a straight line. We will consider the case $N > 1$, too.

The system (1) must be supplemented by initial conditions

$$\begin{aligned} u(x,0) &= u_o(x) \\ v(x,0) &= v_o(x) \end{aligned} \qquad \text{for all } x \in R^N. \tag{1c}$$

Most commonly, for the function $f: v \in R \to f(v) \in R$ there is chosen a cubic polynomial with the zeros 0,1 and $a \in (0,1)$:

$$f(v) = v(1-v)(v-a). \tag{f}$$

In the following we admit more general nonlinearities $f \in C^1(R)$. We list the relevant assumptions on f (Lateron, it will always be stated, which ones of them are needed):

There exists a constant $a_1 \in (0,\infty)$ such that (f1)
$f(v) \leq a_1\, v$ for all $v \in [0,\infty)$,
$f(v) \geq a_1\, v$ for all $v \in (-\infty,0]$.

Define the primitive of f by $F(v) = \int_0^v f(\tilde{v})\, d\tilde{v}$ for all $v \in R$.
Assume that (f1) holds. Define $a_2 \in R$ and $a_3 \in R$ as the maxima of real numbers for which (f2) and (f3) hold, respectively.

$$\sigma v^2/2 - cF(v) \geq a_2 v^2 \quad \text{for all } v \in R \tag{f2}$$

$$\sigma c v^2 - (\sigma + c^2) v f(v) \geq a_3 v^2 \quad \text{for all } v \in R \tag{f3}$$

If (f1) holds for some $a_1 < \sigma/c$, then (f2) holds for some $a_2 > 0$.
If (f1) holds for some $a_1 < \sigma c/(\sigma + c^2)$, then (f2)(f3) hold for some $a_2, a_3 > 0$.

In the following p.148 - 156, we denote by a_i constants depending only on $\sigma, c \in (0,\infty)$ and the nonlinearity $f \in C^1(R)$.
In the following p.149 - 156, we denote by M_i constants depending on $\sigma, c \in (0,\infty)$, the function $f \in C^1(R)$ and exponents.

We denote by $S(t)$ the formal semigroup generated by $-\Delta$ in R^N. More explicitly, the operator $S(t)\colon u \in C(R^N) \to S(t)u \in C(R^N)$ is given by

$$[S(t)u](x) = (4\pi t)^{-N/2} \int_{R^N} \exp -\left[\frac{(x-y)^2}{4t}\right] u(y)\, dy \quad \text{for all } t \in (0,\infty),\ u \in C(R^N). \tag{2}$$

The relevant estimate for this operator is given by (37) p.26.

<u>Theorem 1</u> (Global solutions for arbitrary space dimension)

Let $p_0 \in [1,\infty]$, $\alpha \in (0,1)$ satisfy $n/p_0 < 1$ (Recall that $n = N/2$).
Take initial data $v_0 \in L_{p_0}(R^N) \cap C^{2+\alpha}(R^N)$ and $u_0 \in L_{p_0}(R^N) \cap C^{\alpha}(R^N)$.
Assume that the nonlinear function $f \in C^1(R)$ satisfies the one-sided sublinearity assumption (f1).

Then the FitzHugh-Nagumo system (1a)(1b)(1c) has a global classical solution (u,v) on the time interval $(0,\infty)$. Furtheron, there exists a constant $M_1 \in (0,\infty)$ such that (5)(6) hold for all $t \in (0,\infty)$.

$$\lim_{t\to 0} [\|v(t) - v_0\|_{p_0} + \|u(t) - u_0\|_{p_0}] = 0; \tag{3}$$

$$\lim_{t\to 0} [\|v(t) - S(t)v_0\|_\infty + \|u(t) - u_0\|_\infty] = 0; \tag{4}$$

$$\|v(t)\|_{p_0} + \|u(t)\|_{p_0} \le M_1 e^{M_1 t}(\|v_0\|_{p_0} + \|u_0\|_{p_0}); \tag{5}$$

$$m(t)^{n/p_0}\|v(t)\|_\infty + \|u(t) - e^{-ct}u_0\|_\infty \le M_1 e^{M_1 t}(\|v_0\|_{p_0} + \|u_0\|_{p_0}). \tag{6}$$

(We have dropped ".,", hence $v(.,t) = v(t)$, $u(.,t) = u(t)$)

<u>Remarks</u>

(i) For simplicity, we do not extend this result to arbitrary initial data $u_0, v_0 \in L_{p_0}(R^N)$.

(ii) Under the stronger restriction on the nonlinearity f

$$\lim_{|v|\to\infty} |f(v)/v| > \sigma/c, \tag{f4}$$

there exist arbitrary large invariant rectangles. (This holds e.g. if f is given by the polynomial (f)). Then global uniform bounds and global existence of a solution are easy to show (see Rauch and Smoller [51]).

Proof

Recall the mild formulation of the FitzHugh-Nagumo system (1)

$$v(t) = S(t)v_o + \int_0^t S(t-s)(f(v)-u)(s)\,ds \tag{7a}$$

$$u(t) = e^{-ct}u_o + \int_0^t \sigma e^{-c(t-s)}v(s)\,ds \tag{7b}$$

(We have dropped ., $v(t) = v(.,t)$ and $f(v)(s) = f(v(.,s))$ and $u(t) = u(.,t)$ are of course functions of the space variable $x \in R^N$).

For the construction of a mild solution $w = (v,u)$ we choose the Banach space $X \times X$, where $X = L_{p_o}(R^N) \cap L_\infty(R^N)$ endowed with the norm

$$\|u\|_X = \max\{\|u\|_{p_o}, \|u\|_\infty\}.$$

Define

$$L(U) = \sup\{|f'(u)|; |u| \leq U\} \quad \text{for all } U \in (0,\infty).$$

In the space X, the nonlinear term (the corresponding Nemyzky operator) f satisfies the local Lipschitz condition

$$\|f(u)-f(v)\|_X \leq L(\|u\|_\infty + \|v\|_\infty)\|u-v\|_X \quad \text{for all } u,v \in X. \tag{f5}$$

By the standard Picard-Lindelöf iteration, one proves that in the space $X \times X$ the integral equation (7a)(7b) has unique solution

$$(v,u)\colon t \in [0,T_{max}) \to (v(t),u(t)) \in X \times X.$$

Let $T_{max} \in (0,\infty]$ be chosen maximal. Then one gets

$$\lim_{t\to T_{max}} \|v(t)\|_X + \|u(t)\|_X = \infty \quad \text{if } T_{max} < \infty. \tag{8}$$

Assertions (3)(4) are evident from this construction.

Now assume additionally that the initial data are smooth, let $v_o^o \in C^{2+\alpha}(R^N)$ and $u_o \in C^\alpha(R^N)$. Then the solution constructed above is identical with the well-known classical solution (see Ladyzenskaja [32] p.320, Theorem 5.1). For the classical solution we prove estimates:

Lemma 1 proves L_p-estimates for v and u by an appropriate decomposition $v = v_1 - v_2$, $u = u_1 - u_2$, where v_i, u_i can be shown to be nonnegative.

Lemma 2 gives L_∞-estimates for v and u.

Assuming these Lemmas we finish the proof.

Applying Lemma 1 with $p_o = q_o$ from the Theorem and $p = \infty$ as well as $p = p_o$ yields the estimate

$$\|v(t)\|_X + \|u(t) - e^{-ct}u_o\|_X \leq K\, t^{-n/p_o} e^{Mt} \, (\|v_o\|_{p_o} + \|u_o\|_{p_o}) \quad \text{for all } t \in (0, T_{max}).$$

Thus (8) implies $T_{max} = \infty$ and assertions (5)(6) follow.

Thus the Theorem is proved.

We state and prove the announced Lemmas.

Lemma 1

Let $p_o, q_o \in [1,\infty]$, $p \in [p_o,\infty] \cap [q_o,\infty]$, $\alpha \in (0,1)$ and

$\delta = n/p_o - n/p$, $\eta = n/q_o - n/p < 1$.

Take initial data $v_o \in L_{p_o}(R^N) \cap C^{2+\alpha}(R^N)$ and $u_o \in L_{q_o}(R^N) \cap C^\alpha(R^N)$.

Let the nonlinear term $f \in C^1(R)$ satisfy (f1).

Then there exists $M_2 \in (0,\infty)$ such that the classical solution of the system (1a)(1b)(1c) satisfies the estimate

$$\|v(t)\|_p \leq M_2 e^{M_2 t} (t^{-\delta} \|v_o\|_{p_o} + \|u_o\|_{q_o}) \qquad \text{for all } t \in (0, T_{max}). \quad (9)$$

If $\delta \in [0,1)$ then furtheron

$$\|u(t) - e^{-ct}u_o\|_p \leq M_2 e^{M_2 t} (\|v_o\|_{p_o} + \|u_o\|_{q_o}) \quad \text{for all } t \in (0, T_{max}). \quad (10)$$

Proof

For arbitrary $a \in R$, let $a^+ = \max\{0,a\}$, $a^- = -\min\{0,a\}$ and hence $a = a^+ - a^-$ and $|a| = a^+ + a^-$.

We decompose the solution (v,u) of the FitzHugh-Nagumo system (1) as $v = v_1 - v_2$, $u = u_1 - u_2$, where

$$(v_1,v_2,u_1,u_2): (x,t) \in R^N \times [0,\infty) \to (v_1,v_2,u_1,u_2)(x,t) \in R^4$$

is the solution of the following Cauchy problem:

$$
\begin{aligned}
v_{1t} &= \Delta v_1 + (f(v)/v)v_1 + u_2 \\
v_{2t} &= \Delta v_2 + (f(v)/v)v_2 + u_1 \qquad \text{for all } x \in R^N, \\
u_{1t} &= \sigma v_1 - cu_1 \qquad\qquad\qquad\qquad \text{all } t > 0. \qquad (11)\\
u_{2t} &= \sigma v_2 - cu_2
\end{aligned}
$$

with the initial conditions

$$
\begin{aligned}
v_1(x,0) &= v_o^+(x), \quad v_2(x,0) = v_o^-(x) \\
u_1(x,0) &= u_o^+(x), \quad u_2(x,0) = u_o^-(x)
\end{aligned}
\qquad \text{for all } x \in R^N. \qquad (12)
$$

The maximum principle implies $v_1, v_2, u_1, u_2 \geqq 0$.
By assumption (f1) we know that

$$f(v)/v \leqq a_1 \quad \text{for all } v \in R.$$

Hence we get for the functions $V = v_1+v_2$ and $U = u_1+u_2$ the following differential inequalities and initial conditions:

$$
\begin{aligned}
V_t &\leqq \Delta V + a_1 V + U \\
U_t &= \sigma V - cU
\end{aligned}
\qquad \text{for all } x \in R^N,\ t > 0. \qquad (13)
$$

and

$$
\begin{aligned}
V(x,0) &= |v_o(x)| \\
U(x,0) &= |u_o(x)|
\end{aligned}
\qquad \text{for all } x \in R^N. \qquad (14)
$$

Recall that $S(t)$ denotes the formal semigroup generated by $-\Delta$ in R^N. Define for all $t \in (0,\infty)$ the operators $R(t): C(R^N) \rightarrow C(R^N)$ by setting

$$R(t) = \int_0^t e^{a_1(t-s)} S(t-s) e^{-cs}\, ds. \qquad (15)$$

The mild reformulation of (13)(14) is

$$V(x,t) \leqq e^{a_1 t}[S(t)|v_o|](x) + \int_0^t e^{a_1(t-s)}[S(t-s)U(.,s)](x)\, ds \qquad (16)$$

$$U(x,t) = e^{-ct}|u_o(x)| + \sigma\int_0^t e^{-c(t-s)} V(x,s)\, ds \quad \text{for all } x \in R^N, \text{ all } t > 0. \qquad (17)$$

which implies that for all $x \in R^N$, $t > 0$ we have

$$V(x,t) \leqq e^{a_1 t}[S(t)|v_o|](x) + [R(t)|u_o|](x) + \sigma\int_0^t [R(t-s)V(.,s)](x)\, ds. \qquad (18)$$

Let $p \in [1,\infty]$ and $q \in [1,p]$. By estimate (37) p.26 the operator $S(t)$ can be extended to a continuous operator from $L_q(R^N)$ to $L_p(R^N)$ satisfying

$$\|S(t)u\|_p \le K_{20} t^{-(n/q-n/p)} \|u\|_q \quad \text{for all } t \in (0,\infty),\ u \in L_q(R^N).$$

Since $\eta = n/q_o - n/p \in [0,1)$, the operator $R(t)$ defined by (15) can be extended to a continuous operator from $L_{q_o}(R^N)$ to $L_p(R^N)$ satisfying

$$\|R(t)u\|_p \le M_3 e^{a_1 t} m(t)^{1-\eta} \|u\|_{q_o} \quad \text{for all } t \in [0,\infty),\ u \in L_{q_o}(R^N). \tag{19}$$

Now (18) implies with $\delta = n/p_o - n/p \in [0,\infty)$ that

$$\|V(t)\|_p \le M_4 e^{a_1 t}[t^{-\delta}\|v_o\|_{p_o} + m(t)^{1-\eta}\|u_o\|_{q_o}] + M_4\int_0^t e^{a_1(t-s)}\|V(s)\|_p\, ds$$

for all $t \in (0,T_{max})$.

Hence Gronwall's Lemma implies

$$\|V(t)\|_p \le M_5 e^{M_5 t}[t^{-\delta}\|v_o\|_{p_o} + \|u_o\|_{q_o}] \quad \text{for all } t \in (0,T_{max}).$$

Thus (9) is proved. Since (7b) implies the estimate

$$\|u(t) - e^{-ct}u_o\|_p \le \sigma\int_0^t e^{-c(t-s)}\|v(s)\|_p\, ds,$$

we get (10) from (9) and the additional assumption $\delta < 1$.
Thus the Lemma is proved.

Lemma 2

Let $p_o, q_o \in [1,\infty]$ satisfy

$$0 \le n/p_o - n/q_o < 1 \quad \text{and} \quad n/q_o < 1.$$

Take smooth initial data $v_o \in L_{p_o}(R^N) \cap C^{2+\alpha}(R^N)$, $u_o \in L_{q_o}(R^N) \cap C^{\alpha}(R^N)$. Assume that (f1) holds for the nonlinear term $f \in C^1(R)$.

Then the solution (v,u) of the FitzHugh-Nagumo system (1) satisfies

$$\|v(t)\|_\infty \le M_6 m(t)^{-n/p_o} \sup_{0<s\le t} [\|v(s)\|_{p_o} + \|u(s)\|_{q_o}] \quad \text{for all } t \in (0,T_{max}). \tag{20}$$

If $n/p_o < 1$, then furtheron

$$\|u(t) - e^{-ct}u_o\|_\infty \le M_6 \sup_{0<s\le t} [\|v(s)\|_{p_o} + \|u(s)\|_{q_o}] \quad \text{for all } t \in [0,T_{max}). \tag{21}$$

Proof

For $t \le 1$, apply Lemma 1 with $p = \infty$. For $t > 1$ let $t = 1+t_o$ and take $v_o := v(t_o)$, $u_o := u(t_o)$ as initial data and apply Lemma 1 with $p = \infty$.

The following Theorem gives conditions under which the solution of the FitzHugh-Nagumo system decays for large time. Turning it the other way round: if the FitzHugh-Nagumo system really models the transmission of a nerve pulse, some of the assumptions of Theorem 2 must necessarily be violated.

Theorem 2 (Sufficient conditions for decay of solutions in space dimension $N \leq 3$)

Take space dimension $N = 1,2$ or 3, $\alpha \in (0,1)$ and initial data $v_0 \in L_2(R^N) \cap C^{2+\alpha}(R^N)$ and $u_0 \in L_2(R^N) \cap C^{\alpha}(R^N)$.
Let the nonlinearity $f \in C^1(R)$ satisfy (f1) and (f2)(f3) for some $a_2, a_3 > 0$.

Then the global classical solution (v,u) of the FitzHugh-Nagumo system (1a)(1b)(1c) given in Theorem 1 decays for large time to the zero rest state. Hence no nerve pulse can be transmitted. More precisely, we have

$$\lim_{t\to\infty} [\|v(t)\|_2 + \|v(t)\|_\infty + \|u(t)\|_2 + \|u(t)\|_\infty] = 0. \quad (22)$$

Proof
Existence of a global solution (v,u) of the FitzHugh-Nagumo system is shown in Theorem 1. Define the following functional $\Psi(t)$ of the solution $(v,u) = (v,u)(x,t)$:

$$\Psi(t) = \int_{R^N} [c(\nabla v)^2/2 - cF(v) + \sigma v^2/2 + (cv-u)^2/2](x,t)\, dx \quad \text{for all } t \in [0,\infty).$$

A straightforward computation shows

$$\frac{d\Psi}{dt}(t) = -\int_{R^N} [c(\Delta v + f(v))^2 + (\sigma + c^2)(\nabla v)^2 + 2cu^2](x,t)\, dx + \\ + \int_{R^N} [(\sigma + c^2) v f(v) - \sigma c v^2](x,t)\, dx \quad \text{for all } t \in [0,\infty).$$

Hence the assumptions (f2)(f3) imply that there exists $M_6 \in (0,\infty)$ such that

$$\Psi(t) \geq a_2 \|v(t)\|_2^2/2 + M_6 \|u(t)\|_2^2 \quad \text{and}$$

$$\frac{d\Psi}{dt}(t) \leq -a_3 \|v(t)\|_2^2 \quad \text{for all } t \in [0,\infty).$$

By integration with respect to time t we get for some $M_7, M_8 \in (0,\infty)$

$$\|v(t)\|_2^2 + \|u(t)\|_2^2 + M_7 \int_0^t \|v(s)\|_2^2\, ds \leq M_8\, \Psi(0) \quad \text{for all } t \in [0,\infty). \quad (23)$$

Hence by Lemma 2 there exists $V \in (0,\infty)$ such that

$$\sup_{0 \le t < \infty} [m(t)^{n/2} \|v(t)\|_\infty + \|u(t) - e^{-ct}u_0\|_\infty] \le V. \tag{24}$$

Since $f \in C^1(R)$ and $f(0) = 0$ by (f1), there exists $A \in (0,\infty)$ such that

$$|f(v)| \le A|v| \quad \text{for all } |v| \le V. \tag{25}$$

In the following we have $0 \le s \le s+t < \infty$. We use the following mild reformulation of the FitzHugh-Nagumo system (1̇):

$$v(s+t) = e^{-2ct} S(t) v(s) + \int_0^t e^{-2c(t-\tau)} S(t-\tau)[f(v)+2cv-u](s+\tau)\, d\tau \tag{26}$$

$$u(s+t) = e^{-ct} u(s) + \sigma \int_0^t e^{-c(t-\tau)} v(s+\tau)\, d\tau \quad \text{for all } s,t \in (0,\infty). \tag{27}$$

Define

$$Q(t) = \int_0^t e^{-c\tau} S(\tau) d\tau.$$

Inserting (27) in (26) yields

$$\begin{aligned} v(s+t) &= e^{-2ct} S(t) v(s) - e^{-ct} Q(t) u(s) + \\ &+ \int_0^t e^{-2c(t-\tau)} S(t-\tau)[f(v)+2cv](s+\tau)\, d\tau + \\ &- \sigma \int_0^t e^{-c(t-\tau)} Q(t-\tau) v(s+\tau)\, d\tau \qquad \text{for all } s,t \in (0,\infty). \end{aligned} \tag{28}$$

Clearly

$$\|S(t)u\|_2 \le \|u\|_2 \quad \text{for all } t \in [0,\infty),\ u \in L_2(R^N) \tag{29}$$

implies

$$\|Q(t)u\|_2 \le \|u\|_2/c \quad \text{for all } t \in [0,\infty),\ u \in L_2(R^N). \tag{30}$$

Using (25)(29)(30), furtheron (23) and Schwarz's inequality, equation equation (28) yields the estimate

$$\begin{aligned} \|v(s+t)\|_2 &\le e^{-ct}(\|v(s)\|_2 + \|u(s)\|_2/c) + \\ &+ \int_0^t e^{-c(t-\tau)} \|[\,|f(v)| + (2c+\sigma/c)\,|v|\,](s+\tau)\|_2\, d\tau \le \\ &\le e^{-ct} M_9 \Psi(0) + (A+2c+\sigma/c)/\sqrt{2c} \left[\int_0^t \|v(s+\tau)\|_2^2\, d\tau\right]^{1/2} \\ &\qquad \text{for all } s,t \in (0,\infty). \end{aligned} \tag{31}$$

Equation (27) implies the estimate

$$\|u(s+t)\|_2 \le e^{-ct} \|u(s)\|_2 + \sigma \int_0^t e^{-c(t-\tau)} \|v(s+\tau)\|_2\, d\tau \tag{32}$$

and hence

$$\|u(s+t)\|_2 \leq e^{-ct} M_8 \Psi(0) + \sigma/\sqrt{2c}\left[\int_0^t \|v(s+\tau)\|_2^2 \, d\tau\right]^{1/2} \quad \text{for all } s,t \in (0,\infty). \tag{33}$$

In the estimates (31)(33), the first term on the right-hand side decays for $t \to \infty$ and the second for fixed t and $s \to \infty$ by estimate (23). Hence we have shown

$$\lim_{t\to\infty} [\|v(t)\|_2 + \|u(t)\|_2] = 0. \tag{34}$$

Taking v(s) and u(s) as initial values we get analogously to (20)(21)

$$\|v(s+t)\|_\infty \leq M_{10} m(t)^{-n/2} \sup_{0<\tau<t} [\|v(s+\tau)\|_2 + \|u(s+\tau)\|_2], \tag{35}$$

$$\|u(s+t) - e^{-ct} u(s)\|_\infty \leq M_{10} \sup_{0<\tau<t} [\|v(s+\tau)\|_2 + \|u(s+\tau)\|_2] \quad \text{for all } s,t \in (0,\infty). \tag{36}$$

Now (24)(34)(35)(36) imply

$$\lim_{t\to\infty} [\|v(t)\|_\infty + \|u(t)\|_\infty] = 0. \tag{37}$$

By (34)(37) the assertion (22) is shown. Thus the Theorem is proved.

Chemical Reactions

Let A, B, C be chemicals reacting according to

$$A + B \rightleftharpoons C. \tag{1}$$

Assume that these substances are confined in some bounded domain $\Omega \subset R^N$ with impermeable boundaries where they diffuse with rates $D_a, D_b, D_c > 0$ and react according to (1). Then the concentrations are functions

$$(a,b,c): (x,t) \in \Omega \times [0,\infty) \to (a,b,c)(x,t) \in R^3$$

obeying the following initial-boundary value problem:

$$a_t - D_a \Delta a = c - ab \tag{2a}$$

$$b_t - D_b \Delta b = c - ab \qquad \text{for all } x \in \Omega,\ t > 0; \tag{2b}$$

$$c_t - D_c \Delta c = -c + ab \tag{2c}$$

$$\frac{\partial a}{\partial n} = 0, \quad \frac{\partial b}{\partial n} = 0, \quad \frac{\partial c}{\partial n} = 0 \qquad \text{for all } x \in \partial\Omega,\ t > 0; \tag{2d}$$

$$a(x,0) = a_o(x), \quad b(x,0) = b_o(x), \quad c(x,0) = c_o(x) \quad \text{for all } x \in \Omega \tag{2e}$$

Although the results of the following Theorems 1 and 2 seem quite clear intuitively, the proofs turn out to be quite difficult.

Theorem 1 (Globally bounded solutions for space dimension $N \leq 5$)
Take space dimension $N \leq 5$ and let $p_a, p_b, p_c \in [1,\infty]$ satisfy $(n = N/2)$

$$1/p_a < 1/2 + 1/(2n), \quad 1/p_b < 1/2 + 1/(2n). \tag{3}$$

Assume for the initial data $a_o \in L_{p_a}(\Omega)$, $b_o \in L_{p_b}(\Omega)$, $c_o \in L_{p_c}(\Omega)$ and $a_o, b_o, c_o \geq 0$.

Then the initial-boundary value problem (2a)-(2e) has a global classical solution on the time interval $(0,\infty)$ satisfying the initial condition (2e) in the sense

$$\lim_{t\to 0} [\|a(t)-a_o\|_1 + \|b(t)-b_o\|_1 + \|c(t)-c_o\|_1] = 0. \tag{4}$$

Furtheron there exists an exponent $\bar{\delta} \in [0,\infty)$ depending only on N, p_a, p_b, p_c such that

$$\sup \{m(t)^{\bar{\delta}} [\|a(t)\|_\infty + \|b(t)\|_\infty + \|c(t)\|_\infty];\ t \in (0,\infty)\} < \infty. \tag{5}$$

If $p_a = p_b = p_c = \infty$, then (5) holds with $\overline{\delta} = 0$. In the general case, $\overline{\delta}$ is specified in Lemma 6.
We have dropped ".,", hence $a(t) = a(.,t)$ etc..

Proof
The set of convenient initial data consists of all (a_o, b_o, c_o) such that $a_o, b_o, c_o \geqq 0$ and $a_o \in L_{p_a}$, $b_o \in L_{p_b}$, $c_o \in L_{p_c}$ with p_a, p_b, p_c as above.

In the following Lemmas 1 through 6 we can restrict ourselves to regular convenient initial data (see Definition 1 p.110). By Theorem 1 p.111, the system (2a)-(2e) has a classical solution for regular initial data on some maximal time interval $[0, T_{max})$. The maximum principle p.123 yields for this solution (a,b,c):

$$a(x,t) \geqq 0,\ b(x,t) \geqq 0,\ c(x,t) \geqq 0 \quad \text{for all } x \in \overline{\Omega},\ t \in [0, T_{max}). \tag{6}$$

Denote by $S_a(t), S_b(t), S_c(t)$ the semigroups generated by the operators $-D_a\Delta+1$, $-D_b\Delta+1$, $-D_c\Delta+1$ in the domain Ω with Neumann boundary conditions on $\partial\Omega$. We use the mild reformulation of system (2):

$$a(t) = S_a(t)a_o + \int_0^t S_a(t-s)[a + c - ab](s)\, ds \tag{7a}$$

$$b(t) = S_b(t)b_o + \int_0^t S_b(t-s)[b + c - ab](s)]ds \tag{7b}$$

$$c(t) = S_c(t)c_o + \int_0^t S_c(t-s)[ab](s)\, ds \tag{7c}$$

for all $t \in (0, T_{max})$.

Since by Lemma 2(i) p.19 the semigroups S are positive, (6)(7) imply

$$0 \leqq a(x,t) \leqq S_a(t)a_o + \int_0^t [S_a(t-s)[a+c](.,s)](x)\, ds \tag{8a}$$

$$0 \leqq b(x,t) \leqq S_b(t)b_o + \int_0^t [S_b(t-s)[b+c](.,s)](x)\, ds \tag{8b}$$

$$0 \leqq c(x,t) \leqq S_c(t)c_o + \int_0^t [S_c(t-s)[ab](.,s)](x)\, ds \tag{8c}$$

for all $x \in \overline{\Omega}$, $t \in (0, T_{max})$.

These estimates are the central tool in the following Lemmas.
The scheme depicted in fig.1 gives a survey how estimates of different quantities imply one another. Here we have used the shorthand notation $Sa_o + S*[a+c]$ for the right-hand side of (8a) etc.

Lemma 1 treats the central part of the figure.

Lemma 2 treats the upper and lower parts.

Lemma 3 contains the "feedback"-argument.

Lemma 4 completes the estimates sketched in the figure.

Lemma 5 proves that the initial data are approached for $t \to 0$.

Lemma 6 derives L_∞-estimates by successive application of Lemma 4 for suitable exponents.

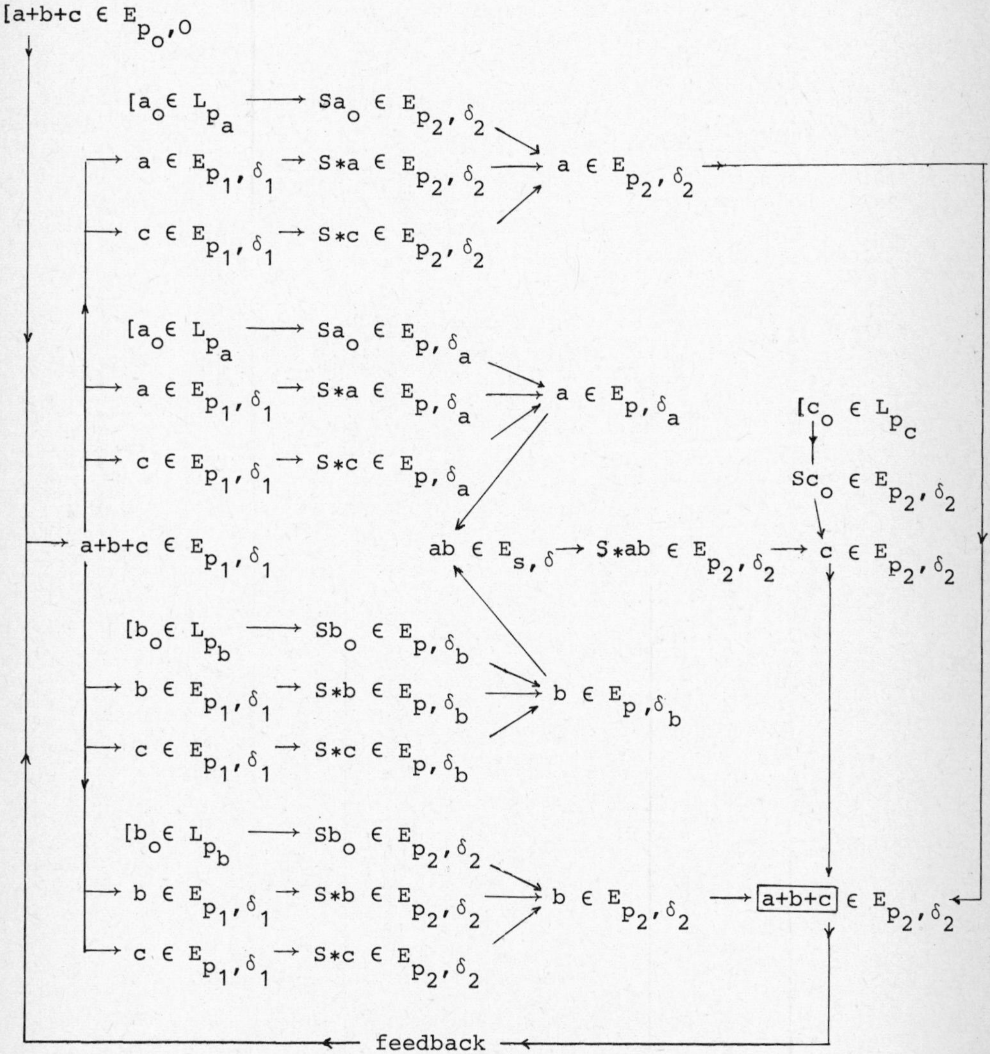

figure 1 inputs are marked by [, the output by ▭

Lemma 1

Let $p_1, p_2, p_a, p_b, p_c \in [1,\infty]$ and $\delta_1, \delta_2 \in [0,1)$ satisfy $(n = N/2)$

$$n/p_1 < n/2 + 1 \ , \quad n/p_1 < n/(2p_2) + 3/2 \ ; \tag{9}$$

$$n/p_a \leq n/2 + \delta_2/2 \ , \quad n/p_a < n/(2p_2) + 1/2 + \delta_2/2 \ ; \tag{10a}$$

$$n/p_b \leq n/2 + \delta_2/2 \ , \quad n/p_b < n/(2p_2) + 1/2 + \delta_2/2 \ ; \tag{10b}$$

$$\delta_1 = \delta_2/2 \ , \quad n/p_c \leq n/p_2 + \delta_2 \ . \tag{11}$$

Take convenient regular initial data.

Then there exists a constant K depending only on the exponents n, p_i, δ_i, the domain Ω and D_a, D_b, D_c such that the solution (a,b,c) of the system (2) satisfies the estimate

$$\|c\|_{p_2,\delta_2,T} \leq K[\|a_o\|^2_{p_a} + \|b_o\|^2_{p_b} + \|c_0\|_{p_c} + (\|a+b+c\|_{p_1,\delta_1,T})^2] \tag{12}$$
for all $T \in (0, T_{max})$.

Proof

By assumptions (9)(10)(11) there exists $p \in [2,\infty]$ such that

$$
\begin{aligned}
&1/p_{a,b} - \delta_2/(2n) < 1/p \\
&1/p_1 \ \ - \ 1/n \ \ < 1/p \\
&1/p < 1/(2p_2) + 1/(2n) \\
&1/p \leq 1/2 \\
&1/p_c \ \ - \delta_2/n \ \ \leq 1/p_2 \ .
\end{aligned}
$$

Define $s = p/2$ and $\delta_a = \delta_b = \delta_1$, $\delta = \delta_2$.

Then the following set of conditions hold:

$$n/p_1 - n/p < 1 \ , \quad \delta_1 < 1 \ , \quad \delta_1 + n/p_1 - n/p \leq 1 + \delta_{a,b} \tag{13}$$

$$2/p \leq 1/s \leq 1 \ , \quad \delta_a + \delta_b = \delta \tag{14}$$

$$n/s - n/p_2 < 1 \ , \quad \delta < 1 \ , \quad \delta + n/s - n/p_2 \leq 1 + \delta_2 \tag{15}$$

$$n/p_a - n/p \leq \delta_a \ , \quad n/p_b - n/p \leq \delta_b \ , \quad n/p_c - n/p_2 \leq \delta_2 \ . \tag{16}$$

These conditions are exactly needed to "read backward" the central part of figure 1. Indeed, we derive from (8) via (30) p.25 and Hölder's inequality:

$$\|c\|_{p_2,\delta_2,T} \leq K_1(\|c_0\|_{p_c} + \|ab\|_{s,\delta,T})$$

$$\|ab\|_{s,\delta,T} \leq K_2 \|a\|_{p,\delta_a,T} \|b\|_{p,\delta_b,T}$$ for all $T\in(0,T_{max})$.

$$\|a\|_{p,\delta_a,T} \leq K_3(\|a_0\|_{p_a} + \|a\|_{p_1,\delta_1,T} + \|c\|_{p_1,\delta_1,T})$$

$$\|b\|_{p,\delta_b,T} \leq K_4(\|b_0\|_{p_b} + \|b\|_{p_1,\delta_1,T} + \|c\|_{p_1,\delta_1,T})$$

Now the assertion (12) is straightforward. Thus the Lemma is proved.

Lemma 2

Let $p_1,p_2,p_a,p_b\in[1,\infty]$ and $\delta_1,\delta_2\in[0,\infty)$ satisfy

$$n/p_1 < n/p_2 + 1 \;, \quad \delta_1 \leq \delta_2 \;, \tag{17}$$

$$n/p_a - n/p_2 \leq \delta_2 \;, \quad n/p_b - n/p_2 \leq \delta_2 \;. \tag{18}$$

Take convenient regular initial data.

Then there exists a constant K_5 depending only on the exponents, the domain Ω and D_a,D_b,D_c such that the solution (a,b,c) of the system (2) satisfies the estimates

$$\|a + b\|_{p_2,\delta_2,T} \leq K_5[\|a_0\|_{p_a}+\|b_0\|_{p_b}+\|c_0\|_{p_c} + \|a + b + c\|_{p_1,\delta_1,T}] \tag{19}$$

$$\int_0^T \|(a + b + c + ab)(s)\|_1 \, ds \leq \tag{20}$$

$$\leq K_5(1 + T)[\|a_0\|_{p_a}+\|b_0\|_{p_b}+\|c_0\|_{p_c} + \|a + b + c\|_{p_1,\delta_1,T}]$$

for all $T\in(0,T_{max})$.

The proof is straightforward by (8) and (30) p.25.

Lemma 3

Let $p_0,p_1,p_2\in(0,\infty]$, $\delta_0,\delta_1,\delta_2\in[0,\infty)$ and $\gamma\in[1,\infty)$ satisfy (21) and (22a) or (21) and (22b).

$$p_0 \leq p_1 \leq p_2 \;, \quad p_0 < p_2 \;, \quad (\gamma-1)/p_0 + 1/p_2 < \gamma/p_1 \tag{21}$$

$$\delta_0/p_1 + \delta_1/p_2 + \delta_2/p_0 \leq \delta_0/p_2 + \delta_1/p_0 + \delta_2/p_1 \tag{22a}$$

$$\delta_0 \leq \delta_2 \quad \text{and} \quad (\gamma-1)\delta_0 + \delta_2 \leq \gamma\delta_1 \tag{22b}$$

Define $\beta = (1/p_0 - 1/p_1)/(1/p_0 - 1/p_2)$

and $\alpha = (1-\beta)\gamma(1-\gamma\beta)^{-1} = (\gamma/p_1 - \gamma/p_2)(\gamma/p_1 - 1/p_2 - (\gamma-1)/p_0)^{-1}$.

Then $\alpha\in[1,\infty)$, $\beta\in[0,1/\gamma)$ and

$$\|c\|_{p_1,\delta_1,T} \leq \|c\|^{1-\beta}_{p_0,\delta_0,T}\,\|c\|^{\beta}_{p_2,\delta_2,T} \quad \text{for all } c \in E_{p_0,\delta_0,T} \cap E_{p_2,\delta_2,T}. \tag{23}$$

Let $\varphi_\alpha: x\in[0,\infty) \to \varphi_\alpha(x)\in[0,\infty)$ be any one of the three functions

$$\varphi_\alpha(x) = x^\alpha \quad \text{or} \quad \varphi_\alpha(x) = \max\{x,x^\alpha\} \quad \text{or} \quad \varphi_\alpha(x) = \max\{1,x^\alpha\}.$$

Then the following implication holds:

$$\begin{aligned} &\text{If } c \in E_{p_2,\delta_2,T} \text{ and } &\|c\|_{p_2,\delta_2,T} &\leq \varphi_\gamma(\|c\|_{p_1,\delta_1,T}),\\ &\text{then} &\|c\|_{p_1,\delta_1,T} &\leq \varphi_{\alpha/\gamma}(\|c\|_{p_0,\delta_0,T})\\ &\text{and} &\|c\|_{p_2,\delta_2,T} &\leq \varphi_\alpha(\|c\|_{p_0,\delta_0,T}). \end{aligned} \tag{24}$$

<u>Proof</u>

In the special case that $\delta_i = 0$ for $i = 0,1,2$ estimate (23) follows by Hölder's inequality. Hence under the assumption

$$\delta_1 \geqq (1-\beta)\delta_0 + \beta\delta_2, \tag{25}$$

estimate (23) holds, too. Assumption (25) is equivalent to (22a). Since $\gamma\beta < 1$, (22b) implies (22a). Hence (23) holds under the assumptions of the Lemma.

It remains to check (24). We consider only the case that $\varphi_\alpha(x) = x^\alpha$. Since we may assume $c \in E_{p_0,\delta_0,T} \cap E_{p_2,\delta_2,T}$, (23) and the first line of (24) imply

$$\|c\|_{p_2,\delta_2,T} \leq \|c\|^{1-\beta}_{p_0,\delta_0,T}\,\|c\|^{\gamma\beta}_{p_2,\delta_2,T}$$

and hence

$$\|c\|_{p_2,\delta_2,T} \leq \|c\|^{(1-\beta)/(1-\gamma\beta)}_{p_o,\delta_o,T} = \varphi_\alpha(\|c\|_{p_0,\delta_0,T})$$

which is the third line of (24). To prove the second line of (24), use (23) once more:

$$\|c\|_{p_1,\delta_1,T} \leq \|c\|^{1-\beta+\alpha\beta}_{p_0,\delta_0,T} = \varphi_{\alpha/\gamma}(\|c\|_{p_0,\delta_0,T}).$$

Hence (24) is shown. The other choices of φ_α follow easily. Thus the Lemma is proved.

<u>Lemma 4</u>

Let $p_o \in (0,\infty)$, $p_2 \in (1,\infty]$, $p_a, p_b, p_c \in [1,\infty]$, $\delta_2 \in [0,1)$ satisfy

$$n/p_o + n/p_2 < 2 + n \ , \quad n/p_o < 3 \ , \quad n/p_o < n/p_2 + 2 \ , \quad p_o < p_2 \ , \tag{26}$$

$$n/p_a \leq n/2 + \delta_2/2 \ , \quad n/p_a < n/(2p_2) + 1/2 + \delta_2/2 \ , \tag{27a}$$

$$n/p_b \leq n/2 + \delta_2/2 \ , \quad n/p_b < n/(2p_2) + 1/2 + \delta_2/2 \ , \tag{27b}$$

$$n/p_a \leq n/p_2 + \delta_2 \ , \quad n/p_b \leq n/p_2 + \delta_2 \ , \quad n/p_c \leq n/p_2 + \delta_2 \ . \tag{28}$$

Take convenient regular initial data. Define

$$A = \|a_o\|_{p_a} + \|b_o\|_{p_b} + \|c_o\|_{p_c} \ , \tag{29}$$

$$A_{o,T} = \|a\|_{p_o,0,T} + \|b\|_{p_o,0,T} + \|c\|_{p_o,0,T}. \tag{30}$$

Then there exists an exponent $\alpha \in [1,\infty)$ depending only on the exponents n, p_o, p_2 as well as a constant K_6 depending only on the exponents, the domain Ω and D_a, D_b, D_c such that the solution (a,b,c) of the system (2) satisfies the estimate

$$\|a + b + c\|_{p_2,\delta_2,T} \leq K_6[A + A_{o,T} + (A + A_{o,T})^\alpha] \tag{31}$$

for all $T \in (0,T_{max})$.

<u>Proof</u>

Under the assumptions of the Lemma, there exists $p_1 \in [1,\infty]$ such that

$$1/p_o + 1/p_2 < 2/p_1$$

$$2/p_1 < 1 + 2/n$$

$$2/p_1 < 1/p_2 + 3/n$$

$$2/p_1 < 2/p_2 + 2/n$$

$$2/p_2 \leq 2/p_1 \leq 2/p_o \ .$$

Hence the assumptions (9)(10a)(10b)(11) of Lemma 1:

$$n/p_1 < n/2 + 1 \ , \quad n/p_1 < n/(2p_2) + 3/2 \ ; \tag{9}$$

$$n/p_{a,b} \leq n/2 + \delta_2/2 \ , \quad n/p_{a,b} < n/(2p_2) + 1/2 + \delta_2/2 \ ; \tag{10a,b}$$

$$\delta_1 = \delta_2/2 \ , \quad n/p_c \leq n/p_2 + \delta_2 \ ; \tag{11}$$

assumptions (17)(18) of Lemma 2:

$$n/p_1 < n/p_2 + 1 \ , \quad \delta_1 \leq \delta_2 \ , \quad n/p_{a,b} \leq n/p_2 + \delta_2 \ ; \tag{17)(18}$$

assumptions (21)(22b) of Lemma 3 with $\delta_o := 0$, $\gamma := 2$:

$$p_o \leq p_1 \leq p_2 , \quad p_o < p_2 , \quad 1/p_o + 1/p_2 < 2/p_1 , \quad \delta_2 \leq 2\delta_1 \qquad (21)(22b)$$

are satisfied.

Estimates (12), Lemma 1 and (19), Lemma 2 yield

$$\|a+b+c\|_{p_2,\delta_2,T} \leq K_7(A + \|a+b+c\|_{p_1,\delta_1,T} + (A + \|a+b+c\|_{p_1,\delta_1,T})^2)$$

for all $T \in (0,T_{max})$.

Hence assertion (24) of Lemma 3 with $\gamma = 2$, $\delta_o = 0$ implies

$$\|a+b+c\|_{p_i,\delta_i,T} \leq K_8(A + A_{o,T} + (A + A_{o,T})^{\alpha i/2}) \quad \text{for } i = 1,2, \text{ all } T \in (0,T_{max}). \qquad (32)$$

Here $\alpha = 2(1/p_1 - 1/p_2)(2/p_1 - 1/p_2 - 1/p_o)^{-1}$ as shown in Lemma 3. Hence the Lemma is proved.

Lemma 5

Under the assumptions of Lemma 4 there exist $\varepsilon \in (0,1)$, $\alpha \in [1,\infty)$ depending only on the exponents and a constant K_9 depending on the exponents, the domain Ω and D_a, D_b, D_c such that the solution (a,b,c) of system (2) approaches the initial data in the sense (A, $A_{o,T}$ are given by (29)(30))

$$\|a(t) - S_a(t)a_o\|_1 + \|b(t) - S_b(t)b_o\|_1 + \|c(t) - S_c(t)c_o\|_1 \leq \qquad (33)$$

$$\leq K_9(t^{\varepsilon} + t)[A + A_{o,t} + (A + A_{o,t})^{\alpha}] \quad \text{for all } t \in (0,T_{max}).$$

Proof

The mild formulation (7) implies that

$$\|a(t)-S_a(t)a_o\|_1 + \|b(t)-S_b(t)b_o\|_1 + \|c(t)-S_c(t)c_o\|_1 \equiv d(t) \leq$$

$$\leq \int_0^t \|(a + b + c + ab)(s)\|_1 \, ds \quad \text{for all } t \in (0,T_{max}).$$

By the proof of Lemma 1 we can estimate with $\varepsilon = 1 - \delta_2$ ($\delta_2 = \delta = 2\delta_1$)

$$d(t) \leq \|a + b + c\|_{p_1,\delta_1,t} \int_0^t m(s)^{-\delta_1} ds + \|ab\|_{s,\delta,t} \int_0^t m(s)^{-\delta} ds \leq$$

$$\leq K_{10}(t^{\varepsilon}+t)[\|a+b+c\|_{p_1,\delta_1,t} + (\|a_o\|_{p_a} + \|a+c\|_{p_1,\delta_1,t})(\|b_o\|_{p_b} + \|b+c\|_{p_1,\delta_1,t})].$$

Hence by (32) from Lemma 4 above we get the assertion, namely

$$d(t) \leq K_{11}(t^{\varepsilon} + t)(A + A_{o,t} + (A + A_{o,t})^{\alpha}) \quad \text{for all } t \in (0,T_{max}).$$

Lemma 6

Take space dimension $N < 6$ and let $p_a, p_b, p_c \in [1,\infty]$ satisfy one of the following three conditions (34a)(34b)(34c) $(n = N/2)$:

$$N = 1,2 \text{ or } 3\ ,\ p_a, p_b \in (1,\infty] \quad \text{and} \quad n/p_a, n/p_b, n/p_c < 1\ . \tag{34a}$$

$$N < 6\ ,\ p_a, p_b \in [2,\infty] \quad \text{and} \quad n/p_a, n/p_b, n/p_c < 1\ . \tag{34b}$$

$$N < 6\ ,\ n/p_a, n/p_b < n/2 + 1\ . \tag{34c}$$

Take convenient regular initial data a_o, b_o, c_o with

$$A = \|a_o\|_{p_a} + \|b_o\|_{p_b} + \|c_o\|_{p_c}.$$

Then there exists an exponent α depending only on N, p_a, p_b, p_c and a constant K_{12} depending on N, p_a, p_b, p_c, the domain Ω and D_a, D_b, D_c such that the solution of system (2) satisfies the estimate

$$\|(a + b + c)(t)\|_\infty \leq K_{12} m(t)^{-\overline{\delta}}(A + A^\alpha) \quad \text{for all } t \in (0, T_{max}). \tag{35}$$

Here $\overline{\delta} \in [0,\infty)$ is given by

$$\overline{\delta} = \begin{cases} \max\{n/p_a, n/p_b, n/p_c\} & \text{in case that (34a) or (34b) holds,} \\ \alpha & \text{in case that (34c) holds.} \end{cases} \tag{36}$$

Proof

Note that the differential equations (2a)(2b)(2c) and the Neumann boundary conditions (2d) imply the conservation laws

$$\|(a + c)(t)\|_1 = \|a_o + c_o\|_1 \quad \text{and} \quad \|(b + c)(t)\|_1 = \|b_o + c_o\|_1 \quad \text{for all } t \in [0, T_{max}). \tag{37}$$

With A and $A_{o,T}$ specified by (29)(30) in Lemma 4 and $p_o = 1$, we get the primary a priori estimate

$$A_{o,T} \leq 2A \quad \text{for all } T \in (0, T_{max}). \tag{38}$$

We distinguish the three cases that assumption (34a),(34b) or (34c) holds.

Assume (34a).

The assumptions of Lemma 4 are satisfied for

$$N \leq 3,\ p_o := 1,\ p_2 := \infty,\ p_a, p_b, p_c \text{ as above and } \delta_2 := \overline{\delta} \text{ from (36).}$$

Hence assertion (31) of Lemma 4 proves the assertion (35).

Assume (34b).

There exists $p_2 \in (1,\infty)$ such that

$$n - 2 < n/p_2 < 2/n \,, \quad 2n/p_{a,b} - 1 < n/p_2 \,, \quad n/p_{a,b,c} \le n/p_2 \,.$$

Thus the assumptions of Lemma 4 are satisfied with

$$N < 6,\ p_o := 1,\ p_a, p_b, p_c \text{ from above and } \delta_2 := 0.$$

Hence assertion (31) of Lemma 4 yields

$$A_{2,T} \equiv \| a + b + c \|_{p_2,0,T} \le K_{13}(A + A_{o,T} + (A + A_{o,T})^{\alpha}) \qquad (39)$$

for all $T \in (0,T_{max})$.

Hence the primary a priori estimate (38) implies

$$A_{2,T} \le K_{14}\, A \qquad \text{for all } T \in (0,T_{max}). \qquad (40)$$

In a second step we apply Lemma 4 once more (with the initial time 0) and choosing

$$p_o := p_2,\ p_2 := \infty,\ \delta_2 := \bar{\delta} \text{ from (36) and } p_{a,b,c} \text{ as above.}$$

Assertion (31) of Lemma 4 yields

$$\| a + b + c \|_{\infty,\delta,T} \le K_{15}(A + A_{2,T} + (A + A_{2,T})^{\alpha}) \qquad (41)$$

for all $T \in (0,T_{max})$.

Now (40)(41) imply the assertion (35).

Assume (34c).

We construct a finite sequence of exponents

$$1 = \bar{p}_o < \bar{p}_2 < \bar{p}_3 < \ldots < \bar{p}_k = \infty$$

such that the assumptions (26)(27a)(27b)(28) of Lemma 4 are satisfied in the following cases 1),2),3)...k-1):

1) With p_a, p_b, p_c as assumed in (34c) above, $p_o := 1$, $p_2 := \bar{p}_2$ from the sequence above and some suitable $\delta_2 \in [0,1)$.

i) for i = 2...k-1:
With p_a, p_b, p_c redefined by $p_a := p_b := p_c := p_o := \bar{p}_i$ from the sequence above, p_2 redefined by $p_2 := \bar{p}_{i+1}$ from the sequence above and some suitable $\delta_2 \in [0,1)$.

We sketch the construction of the sequence $\bar{p}_i$:

For space dimension $N = 1$, let $k = 2$, $\bar{p}_o = 1$, $\bar{p}_2 = \infty$.

For space dimension $N \in \{2,3,4,5\}$, let $k \le 6$ and choose $\bar{p}_i$ such that

$n/\overline{p}_i + n/\overline{p}_{i+1} < n+2,\ n/\overline{p}_i < n/2+1/2,\ n/\overline{p}_i < n/(2\overline{p}_{i+1})+1$ for $i=2\ldots k-1$.
(To this end, let $n/\overline{p}_2 \simeq n/2+1/2$ and $n/\overline{p}_i \simeq n/(2\overline{p}_{i+1})+1$ for $i=2\ldots k-2$, $n/\overline{p}_{k-1} < 1$)
Some calculations show that $(\overline{p}_i)_{i=1\ldots k}$ is a suitable sequence.

For $i = 2\ldots k$ choose $t_i \in (0,\infty)$ such that $(t_2+\ldots+t_i, T) \neq \emptyset$ and define

$$A_i = \sup\{\|(a+b+c)(s)\|_{\overline{p}_i} \mid s \in [t_2+\ldots+t_i, T]\}.$$

Applying Lemma 4 with the exponents from 1), assertion (31) yields

$$A_2 \leq K_{16} m(t_2)^{-\delta_2}(A + A^{\alpha}) \qquad \text{for all } t_2 \in (0,\infty),\ T \in (0,T_{max}). \quad (42)$$

Now successively for $i = 2\ldots k-1$, we consider $(a,b,c)(t_2+\ldots+t_i)$ as in initial data for an initial-boundary value problem with $t = t_2+\ldots+t_i$.
Applying Lemma 4 with the exponents from i), assertion (31) yields

$$A_{i+1} = K_{17}^{(i)} m(t_{i+1})^{-\delta_2^{(i)}}(A_i + A_i^{\alpha^{(i)}}) \quad \text{for } i = 2\ldots k-1,\ T \in (0,T_{max}), \quad t_2, t_3 \ldots t_{i+1} \in (0,\infty). \quad (43)$$

By a straightforward calculation (42)(43) imply the assertion (41).
Thus the Lemma is proved.

Now the proof of the Theorem is finished by the process outlined p.119 through 123. Arbitrary convenient initial data (a_o, b_o, c_o) can be approximated by a sequence (a_{om}, b_{om}, c_{om}) of regular convenient initial data in the sense of (D) p.119. Then the assertions of the Theorem follow by the arguments of Theorem 2 p.120. Thus the proof is finished.

<u>Theorem 2</u> (Asymptotic behavior by means of entropy)

Make the assumptions of Theorem 1 and (to exclude degenerate cases) let

$$\|a_o\|_1 + \|c_o\|_1 > 0 \quad \text{and} \quad \|b_o\|_1 + \|c_o\|_1 > 0. \quad (44)$$

There exists an equilibrium state $(\overline{a}, \overline{b}, \overline{c}) \in (0,\infty)^3$ uniquely determined by

$$\|a_o\|_1 + \|c_o\|_1 = \overline{a} + \overline{c},\ \|b_o\|_1 + \|c_o\|_1 = \overline{b} + \overline{c} \quad \text{and} \quad \overline{a}\overline{b} = \overline{c}. \quad (45)$$

The global classical solution (a,b,c) of the initial-boundary value problem (2a)-(2e) given by Theorem 1 converges to the equilibrium:

$$\lim_{t\to\infty} [\|a(t) - \overline{a}\|_{C^2} + \|b(t) - \overline{b}\|_{C^2} + \|c(t) - \overline{c}\|_{C^2}] = 0. \quad (46)$$

Proof

The maximum principle implies (6) p.158 and hence by assumption (44), the strong maximum principle implies even

$$a(x,t) > 0,\ b(x,t) > 0,\ c(x,t) > 0 \quad \text{for all } (x,t) \in \overline{\Omega}\times(0,\infty). \tag{47}$$

Define the function $s\colon (a,\overline{a})\in(0,\infty)\times(0,\infty) \to s(a,\overline{a})\in(-\infty,0]$ by setting

$$s(a,\overline{a}) = a - \overline{a} + a(\log \overline{a} - \log a).$$

Existence and uniqueness of the equilibrium $(\overline{a},\overline{b},\overline{c})$ satisfying (45) is straightforward. Furtheron let $(\overline{a},\overline{b},\overline{c}) \in (0,\infty)^3$ be the equilibrium. Define the entropy functional

$$\mathcal{S}\colon v = (a,b,c) \in C(\overline{\Omega},(0,\infty)^3) \to \mathcal{S}(v) \in (-\infty,0]$$

by

$$\mathcal{S}(v) = \int_\Omega [s(a(x),\overline{a}) + s(b(x),\overline{b}) + s(c(x),\overline{c})]dx. \tag{48}$$

The entropy production functional is

$$\mathcal{W}\colon v = (a,b,c) \in C^1(\overline{\Omega},(0,\infty)^3) \to \mathcal{W}(v) \in [0,\infty)$$

with

$$\mathcal{W}(v) = \int_\Omega [(ab-c)(\log ab - \log c)](x,t)dx + \tag{49}$$

$$+ \int_\Omega [D_a(\nabla a)^2/a + D_b(\nabla b)^2/b + D_c(\nabla c)^2/c](x,t)dx.$$

The convexity of the function $s(.,\overline{a})\colon a\in(0,\infty) \to s(a,\overline{a})\in(-\infty,0]$ and some elementary considerations yield the following implications:

$$\mathcal{S}(v) = 0 \quad \text{implies} \quad v = (\overline{a},\overline{b},\overline{c}). \tag{50}$$

$$\mathcal{W}(v) = 0 \quad \text{implies} \quad v = (A,B,C) \text{ with } AB = C \text{ and } \nabla A=\nabla B=\nabla C = 0. \tag{51}$$

Let $(a,b,c) = (a,b,c)(x,t)$ be the classical solution of system (2) given by Theorem 1. Differentiating by the time t yields

$$\frac{d}{dt}\mathcal{S}((a,b,c)(.,t)) = \mathcal{W}((a,b,c)(.,t)) \quad \text{for all } t\in(0,\infty) \tag{52}$$

and hence

$$\mathcal{S}((a,b,c)(.,t_2)) - \mathcal{S}((a,b,c)(.,t_1)) = \int_{t_1}^{t_2} \mathcal{W}((a,b,c)(.,s))\,ds \tag{53}$$

for all $t_1,t_2\in(0,\infty)$.

We consider the asymptotic behavior for $t \to \infty$. By (53) the limes

$$\lim_{t\to\infty} \mathcal{S}((a,b,c)(.,t)) = \mathcal{S}_\infty \text{ exists}, \tag{54}$$

$$\int_1^\infty \mathscr{W}((a,b,c)(.,s))\,ds < \infty \quad \text{and} \quad \lim_{t\to\infty}\int_t^\infty \mathscr{W}((a,b,c)(.,s))\,ds = 0. \tag{55}$$

Let $T\in(0,\infty)$ be arbitrary and $\nu\in(1,2)$, $\mu\in(0,1)$ such that $\nu/2+\mu<1$. Define the Banach space

$$X = X^{\nu,\mu} = C^{\mu}([-T,T],C^{\nu}(\overline{\Omega},R^3)).$$

The solution $(a,b,c) = (a,b,c)(x,t)$ of system (2) can be given by the following function

$$U:\ t\in(T,\infty) \to U(t)\in X,$$

$$U(t)(x,s) = (a,b,c)(x,t+s) \quad \text{for all } x\in\overline{\Omega}, |s|\le T.$$

By estimate (5) of Theorem 1 and (262) of Lemma 21 p.78, there exists $M\in(0,\infty)$ such that

$$\|U(t)\|_X \le M \quad \text{for all } t\in[T+1,\infty). \tag{56}$$

Since for $\tilde\nu\in(0,\nu)$, $\tilde\mu\in(0,\mu)$ the embedding $X^{\mu,\nu}\subset X^{\tilde\mu,\tilde\nu}$ is compact, estimate (56) implies that the limit set

$$\omega = \{U\in X \mid \text{there exists a sequence } t_m\to\infty \text{ such that } \lim_{m\to\infty}\|U(t_m)-U\|_X = 0.\} \tag{57}$$

is nonvoid.

Next we show that the limit set ω consists only of one point, namely the equilibrium state $(\overline{a},\overline{b},\overline{c})$ (considered as constant functions). To see this, let $U\in\omega$, $U = (A,B,C)(x,s)$. Then there exists a sequence (t_m) such that

$$\lim_{m\to\infty} t_m = \infty \quad \text{and} \tag{58}$$

$$\lim_{m\to\infty}\max_{|s|\le T}\|a(.,t_m+s)-A(.,s)\|_{C^1(\overline{\Omega})} = 0 \tag{59a}$$

$$\lim_{m\to\infty}\max_{|s|\le T}\|b(.,t_m+s)-B(.,s)\|_{C^1(\overline{\Omega})} = 0 \tag{59b}$$

$$\lim_{m\to\infty}\max_{|s|\le T}\|c(.,t_m+s)-C(.,s)\|_{C^1(\overline{\Omega})} = 0. \tag{59c}$$

Hence (47) and the conservation laws (37) together with (45) imply

$$A(x,s)\ge 0,\ B(x,s)\ge 0,\ C(x,s)\ge 0 \quad \text{for all } x\in\overline{\Omega},\ |s|\le T, \tag{60}$$

$$\|A(.,s)\|_1+\|C(.,s)\|_1 = \overline{a}+\overline{c} \text{ and } \|B(.,s)\|_1+\|C(.,s)\|_1 = \overline{b}+\overline{c} \quad \text{for all } |s|\le T. \tag{61}$$

Since by (55)

$$\lim_{m\to\infty}\int_{-T}^{+T}[(\nabla a)^2/a + (\nabla b)^2/b + (\nabla c)^2/c](.,t_m+s)\,ds = 0,$$

(59)(60) imply $\nabla A = \nabla B = \nabla C = 0$.

(Note at first that $\nabla A(x,s) = 0$ for all $(x,s)\in\overline{\Omega}\times[-T,T]$ such that $A(x,s) > 0$). Hence we get

$$A(x,s) = A(.,s),\ B(x,s) = B(.,s),\ C(x,s) = C(.,s) \quad \text{for all } x \in \overline{\Omega},\ |s| \leq T. \tag{62}$$

We distinguish the cases (a) and (b):

(a) There exists $\hat{s}\in[-T,T]$ such that $C(.,\hat{s}) > 0$.

(b) $C(x,s) = 0$ for all $x \in \overline{\Omega}$, $|s| \leq T$.

Clearly (55) implies

$$\lim_{m\to\infty} \int_{-T}^{+T}\int_{\Omega} [(ab-c)(\log ab - \log c)](x,t_m+s)\, dxds = 0. \tag{63}$$

Consider at first the regular case (a).

Since C is continuous there exists $\sigma > 0$ such that

$$C(.,s) > 0 \quad \text{for all } s \in [\hat{s}-\sigma,\hat{s}+\sigma].$$

We use the convention $(ab-c)(\log ab - \log c) = +\infty$ for $ab = 0$, $c > 0$.
By (59)(63) and the Lemma of Fatou (Segal and Kunze [61] p.69) we get

$$\int_{\hat{s}-\sigma}^{\hat{s}+\sigma} [(AB-C)(\log AB - \log C)](.,s)\, ds = 0.$$

Thus we have shown that

$$(AB-C)(.,s) = 0 \quad \text{for all } s\in[-T,T] \text{ such that } C(.,s) > 0. \tag{64}$$

Now (60)(61)(62)(64) together determine (A,B,C) uniquely. Hence we get

$$A(.,s) = \bar{a},\ B(.,s) = \bar{b},\ C(.,s) = \bar{c} \quad \text{for all } s\in[-T,T] \text{ with } C(.,s) > 0.$$

Since $\bar{c} > 0$ and the function $s \to C(.,s)$ is continuous this implies

$$A(x,s) = \bar{a},\ B(x,s) = \bar{b},\ C(x,s) = \bar{c} \quad \text{for all } x \in \overline{\Omega},\ |s| \leq T. \tag{65}$$

It remains to prove that the exceptional case (b) cannot occur.
To this end, assume that (b) holds. (59)(63) imply

$$AB(.,s) = 0 \qquad \text{for all } s\in[-T,T]. \qquad ?(66)$$

On the other hand, (61)(62) and assumption (b) imply

$$A(.,s) > 0 \text{ and } B(.,s) > 0 \quad \text{for all } s\in[-T,T]. \qquad ?(67)$$

From ?(66) and ?(67) we get a contradiction. Hence the exceptional case (b) cannot occur.

Hence we have proved (65). Thus the limit set consists of a single point:

$$\omega = \{\overline{U}\} \quad \text{with } \overline{U} = (\overline{a},\overline{b},\overline{c}). \tag{68}$$

Now (68) and the definition of the limit set (57) imply

$$\lim_{t\to\infty} \|U(t) - \overline{U}\|_X = 0. \tag{69}$$

For $t\in(T,\infty)$, denote the right-hand side of the differential equations (2a)(2b) by $R(t) \in X$. Hence

$$R(t)(x,s) = (c-ab)(x,t+s) \quad \text{for all } t\in(T,\infty),\ x \in \overline{\Omega},\ |s| \leq T. \tag{70}$$

The mild formulation of the initial-boundary value problem (2) and (69) imply

$$\lim_{t\to\infty} \|R(t)\|_X = 0. \tag{71}$$

Let $\alpha\in(0,1)$ such that $\alpha = \min\{\nu,\mu\}$. Denote by Y the Banach space $Y = C^{2+\alpha,1+\alpha/2}([-T,T]\times\overline{\Omega},R^3)$. The classical Schauder estimates for **linear** parabolic equations (Ladyzenskaja [32], p.320) imply

$$\|U(t) - \overline{U}\|_Y \leq K(T)\|R(t)\|_X \quad \text{for all } t\in(T,\infty). \tag{72}$$

Now (71)(72) yield the assertion (46).

Thus the Theorem is proved.

A Nuclear Reactor Model

The following parabolic system was proprosed by Kastenberg et.al. [26,58] as a simple model for a nuclear reactor:

$$u_t - \Delta u = u(\lambda - bv) \quad \text{for all } x \in \Omega,\ t > 0; \tag{1a}$$

$$v_t = -cv + au \tag{1b}$$

$$u = 0 \quad \text{for all } x \in \partial\Omega,\ t > 0; \tag{1c}$$

$$u = u_o \text{ and } v = v_o \quad \text{for all } x \in \Omega. \tag{1d}$$

Here u represents the fast neutron density and v the fuel temperature (see [26] for the underlying phenomenology).

To give formal definitions, $\Omega \subset R^N$ is a bounded domain with smooth boundary $\partial\Omega \in C^{2+\alpha}$ for some $\alpha \in (0,1)$. Furtheron we have functions $u: (x,t) \in \bar{\Omega} \times [0,\infty) \to u(x,t) \in R$ and $v: (x,t) \in \bar{\Omega} \times [0,\infty) \to v(x,t) \in R$ as well as $u_o: x \in \bar{\Omega} \to u_o(x) \in R$ and $v_o: x \in \bar{\Omega} \to v_o(x) \in R$. $a,b,c,\lambda \in (0,\infty)$ are positive constants in the model equations.

The study of the system (1) begins with the equilibrium states.
Let λ_o be the principle eigenvalue of $-\Delta$ in the domain Ω with Dirichlet boundary conditions on $\partial\Omega$. Let $\varphi_o \in C^2(\bar{\Omega})$ be the corresponding nonnegative eigenfunction:

$$\begin{aligned} -\Delta\varphi_o &= \lambda_o\varphi_o \quad \text{for all } x \in \Omega, \\ \varphi_o &= 0 \quad \text{for all } x \in \partial\Omega. \end{aligned} \tag{2}$$

Existence of the principal eigenfunction φ_o is e.g. proved in Amann [4]. It is well known that $\lambda_o > 0$ and $\varphi_o(x) > 0$ for all $x \in \Omega$.

The nonnegative equilibrium solutions $(\bar{u},\bar{v})$ of the system (1a)(1b)(1c) are given by the solutions of the scalar problem

$$\begin{aligned} -\Delta\bar{u} &= \bar{u}(\lambda - (ab/c)\bar{u}) \quad \text{for all } x \in \Omega, \\ \bar{u} &= 0 \quad \text{for all } x \in \partial\Omega \end{aligned} \tag{3}$$

and $\quad \bar{v} = (a/c)\bar{u}$.

For $\lambda > \lambda_o$, it is straightforward to show that $\varepsilon\varphi_o$ for $\varepsilon > 0$ small enough and the constant M for $M \in (0,\infty)$ large enough are sub- and supersolutions of system (3).

Hence by monotone iteration, it is straightforward that system (3) has a nontrivial, nonnegative solution.

For $\lambda \leq \lambda_o$, system (2) has no nontrivial nonnegative solution.
Define

$$(\bar{u},\bar{v}) = \begin{cases} (0,0) & \text{for } \lambda \leq \lambda_o, \\ (\bar{u},(a/c)\bar{u}) & \text{for } \lambda > \lambda_o. \end{cases} \tag{4}$$

Mottoni and Tesei [42] prove the local stability of the equilibrium $(\bar{u},\bar{v})$ by linearization methods.
C.V.Pao [45,46] considers the system (1) in the case of negative feedback i.e. $b > 0$ - as we do - as well as in the case of positive feedback i.e. $b < 0$ - which makes no sense in the model.
By comparison methods, Pao shows in the case of positive feedback i.e. $a > 0$, $b < 0$ a blow-up in finite time for $\lambda > \lambda_o$ and arbitrary initial data $u_o, v_o > 0$ and for $\lambda < \lambda_o$ and initial data $u_o, v_o > 0$ large enough.
In the more interesting case of negative feedback i.e. $a > 0$, $b > 0$ only the less important subcase $\lambda < \lambda_o$ can be handled by the comparison technique. Theorem 5 in [45] treating the really interesting case $a > 0$, $b > 0$, $c = 0$ and $\lambda > \lambda_o$ is void, because the assumptions are wrong.

In the following we get global results for this interesting case.

<u>Theorem</u> (Boundedness and convergence to equilibrium)

Let space dimension N be arbitrary, $n = N/2$.
Let $p_o \in [1,\infty]$, $q_o \in [2,\infty]$, $q \in [2,q_o]$ and $\nu \in (0,2)$ satisfy

$$n/p_o < 2 \ , \quad n/p_o < n/2 + 1 \ , \tag{5}$$

$$n/q < 1 - \nu/2 \ , \quad n/p_o < n/q + 1 \ . \tag{6}$$

Take initial data $u_o \in L_{p_o}(\Omega)$, $v_o \in L_{q_o}(\Omega)$ such that $u_o \geqq 0$, $v_o \geqq 0$ and furtheron $u_o \in C_o(\bar{\Omega})$ if $p_o = \infty$; $v_o \in C(\bar{\Omega})$, $v_o = \lambda/b$ on $\partial\Omega$ if $q = \infty$;

$$\int_\Omega \bar{u}^{-2} \, |\log(u_o/\bar{u})| \, dx < \infty \quad \text{if } \lambda > \lambda_o. \tag{7}$$

Then the system (1a)-(1d) has a unique global L_1-mild solution (u,v) on the time interval $(0,\infty)$. More precisely, (36a) (36b) p.180 hold with absolutely converging Bochner integrals in $L_1(\Omega)$.
Under the additional assumptions $u_o \in C^{2+\alpha}(\bar{\Omega})$, $v_o \in C^{\alpha}(\bar{\Omega})$ and (11), the functions (u,v) are a classical solution of (1a) (1b) (1c) for $t \in (0,\infty)$.

Furtheron the solution (u,v) satisfies

$$\lim_{t\to 0} \|u(t) - u_o\|_{p_o} = 0; \tag{8a}$$

$$\lim_{t\to 0} \|v(t) - v_o\|_q = 0; \tag{8b}$$

$$\sup_{0<t<\infty} m(t)^{\delta}\|u(t)\|_p < \infty \quad \text{for all } p\in[p_o,\infty],\ \delta = n/p_o - n/p; \tag{9a}$$

$$\sup_{0<t<\infty} \|v(t)\|_q < \infty. \tag{9b}$$

For large t the solution converges to the equilibrium $(\bar{u},\bar{v})$ specified by (4) in the following sense:

$$\lim_{t\to\infty} \|u(t) - \bar{u}\|_{C^{\nu}(\bar{\Omega})} = 0; \tag{10a}$$

$$\lim_{t\to\infty} \|v(t) - \bar{v}\|_q = 0. \tag{10b}$$

<u>Proof</u>

The set $\mathcal{K}$ of convenient initial data consists of all $(u_o,v_o) \in L_{p_o}(\Omega) \times L_{q_o}(\Omega)$ as specified in the Theorem.

The set $\mathcal{R}$ of regular initial data consists of all $(u_o,v_o) \in C^{2+\alpha}(\bar{\Omega}) \times C^{\alpha}(\bar{\Omega})$ such that the following compatibility conditions hold on the boundary:

$$u_o = \Delta u_o = 0 \quad \text{and} \quad \lambda - bv_o = 0 \quad \text{on } \partial\Omega. \tag{11}$$

(Indeed we make this assumption to assure compatibility conditions of first order not only for the system (1), but for the approximating systems (17) for all $\varepsilon\in[0,1)$, too).

By Theorem 1 p.111, the system (1a)(1b)(1c)(1d) has a unique classical solution (u,v) for initial data $(u_o,v_o) \in \mathcal{K} \cap \mathcal{R}$ on some maximal time interval $[0,T_{max})$

Denote by S(t) the semigroup generated by $-\Delta$ in the domain Ω with Dirichlet boundary conditions on $\partial\Omega$. The Comparison Theorem p.123 yields

$$u(x,t) > 0 \quad \text{and} \quad v(x,t) > 0 \quad \text{for all } x \in \Omega,\ t\in(0,T_{max}) \tag{12}$$

and

$$u(x,t) \leq e^{\lambda t}[S(t)u_o](x),\quad v(x,t) \leq a\int_0^t e^{\lambda\tau}[S(\tau)u_o](x)\,d\tau$$
$$\text{for all } x \in \Omega,\ t\in(0,T_{max}).$$

Hence the explosion property (8) of Theorem 1 p.111 shows that $T_{max} = \infty$.

In the following Lemmas 1 through 5 we restrict ourselves to regular convenient initial data.

Lemma 1 constructs a Lyapunov functional. The principal idea is taken from Kawasaki and Teramoto [27]. Formal difficulties arise because of the Dirichlet boundary condition. To overcome them, we consider a sequence of approximating problems.

Lemma 2 uses the results of part I (Lemma 20) to derive L_∞-estimates from the primary a priori estimate given by the Lyapunov functional.

Lemma 3 estimates how the initial data are approximated for $t \to 0$.

Lemma 4 proves continuous dependence of the solution on the initial data on a small time interval.

Lemma 5 shows continuous dependence of the solution on the initial data globally in time.

Lemma 6 gives a weak result about the asymptotic behavior of the component v for $t \to \infty$.

Lemma 7 proves convergence of the solution to the equilibrium for $t \to \infty$.

In Lemma 2 through 7, we denote by the letter M constants depending only on exponents, the domain Ω, the constants a,b,c,λ from the system (1), the equilibrium solution $\bar{u}$ of system (3), the norms $\|u_0\|_{p_0}$, $\|v_0\|_{q_0}$ and the initial value $\Lambda(0)$ of the Lyapunov functional.

We define the following Lyapunov functionals

$$\Lambda: (u,v) \in C(\bar{\Omega},[0,\infty)^2) \to \Lambda(u,v) \in [0,\infty] :$$

In the subcritical case $\lambda \leq \lambda_0$, let

$$\Lambda(u,v) = \int_\Omega \varphi_0(au(x) + bv^2(x)/2)\, dx. \tag{13}$$

In the supercritical case $\lambda > \lambda_0$, let

$$\Lambda(u,v) = \int_\Omega \bar{u}[a(u-\bar{u} - \bar{u}\log(u/\bar{u})) + b(v-\bar{v})^2/2](x)\, dx. \tag{14}$$

<u>Lemma 1</u>

Take regular convenient initial data and let (u,v) be the classical solution of system (1a)-(1d).

Then the function $\Lambda: t\in[0,\infty) \to \Lambda(t)\in[0,\infty)$ is well defined by

$$\Lambda(t) = \Lambda(u(.,t),v(.,t))$$

and satisfies the following equality or inequality, respectively.

In the subcritical case $\lambda \leq \lambda_o$:

$$\Lambda(t_2) + \int_{t_1}^{t_2}\int_\Omega \varphi_o[a(\lambda-\lambda_o)u(x,\tau) + bcv^2(x,\tau)]\, dxd\tau \tag{15}$$
$$= \Lambda(t_1) \qquad \text{for all } t_1,t_2 \text{ with } 0 \leq t_1 < t_2 < \infty.$$

In the supercritical case $\lambda > \lambda_o$:

$$\Lambda(t_2) + \int_{t_1}^{t_2}\int_\Omega a\bar{u}^2[\nabla\log(u(x,\tau)/\bar{u})]^2 + bc\bar{u}[v(x,\tau)-\bar{v}]^2\, dxd\tau \tag{16}$$
$$\leq \Lambda(t_1) \qquad \text{for all } t_1,t_2 \text{ with } 0 \leq t_1 < t_2 < \infty.$$

Proof

To show identity (15) is a straightforward calculation.

We have to show (16). Assume $\lambda > \lambda_o$.
There arise some difficulties because of the Dirichlet boundary condition (1c) and the terms $\log(u/\bar{u})$ and $\nabla\log(u/\bar{u})$ in (16). Thus we consider approximating problems.
Let $\varepsilon\in(0,1)$ and define $(u_\varepsilon,v_\varepsilon)$ as the solution of

$$u_{\varepsilon t} - \Delta u_\varepsilon = (u_\varepsilon + \varepsilon)(\lambda - bv_\varepsilon) \tag{17a}$$
$$\text{for all } x \in \bar{\Omega},\ t\in[0,\infty);$$
$$v_{\varepsilon t} = -cv_\varepsilon + a(u_\varepsilon + \varepsilon) \tag{17b}$$

$$u_\varepsilon = 0 \qquad \text{for all } x \in \partial\Omega,\ t\in[0,\infty); \tag{17c}$$

$$u_\varepsilon = u_o \text{ and } v_\varepsilon = v_o \qquad \text{for all } x \in \bar{\Omega}. \tag{17d}$$

The initial-boundary value problem (17a)-(17d) has a unique classical solution on some maximal time interval $[0,T_\varepsilon)$, since the compatibility condition of first order is satisfied (see Theorem 1 p.111, assumption (S1b) p.110). By comparison arguments, it is easy to see that $u_\varepsilon,v_\varepsilon$ can grow at most exponentially. Hence (8) in Theorem 1 p.112 implies $T_\varepsilon= \infty$. Furtheron the maximum principle implies

$$u_\varepsilon + \varepsilon > 0 \text{ and } v_\varepsilon \geq 0 \text{ for all } x \in \bar{\Omega},\ t\in[0,\infty). \tag{18}$$

Note that in (17) as well as (18) the boundary point $x\in\partial\Omega$ and the case $t = 0$ are included, which is indeed necessary for the calculations below.

For the approximating equations we introduce the functionals

$$\Lambda_\varepsilon(t) = \int_\Omega \bar{u}[a[u_\varepsilon(x,t)+\varepsilon-\bar{u} - \bar{u}\log((u_\varepsilon(x,t)+\varepsilon)/\bar{u})] + b(v_\varepsilon(x,t)-\bar{v})^2/2]\, dx. \tag{19}$$

Differentiating Λ_ε with respect to time t and (3)(17) yield

$$\frac{d\Lambda_\varepsilon}{dt} = \int_\Omega a\overline{u}\left[1 - \frac{\overline{u}}{u_\varepsilon+\varepsilon}\right]\left[\Delta u_\varepsilon + (u_\varepsilon+\varepsilon)(\lambda - bv_\varepsilon)\right] dx +$$

$$+ \int_\Omega b\overline{u}(v_\varepsilon - \overline{v})(-cv_\varepsilon + a(u_\varepsilon+\varepsilon))\, dx$$

$$= \int_\Omega a\overline{u}[u_\varepsilon+ \varepsilon - \overline{u}]\left[\frac{\Delta u_\varepsilon}{u_\varepsilon+\varepsilon} - \frac{\Delta\overline{u}}{\overline{u}}\right] dx +$$

$$+ \int_\Omega a\overline{u}[u_\varepsilon+\varepsilon - \overline{u}][(\lambda - bv_\varepsilon) - (\lambda - b\overline{v})]\, dx +$$

$$+ \int_\Omega b\overline{u}[v_\varepsilon - \overline{v}][-c(v_\varepsilon-\overline{v}) + a(u_\varepsilon+\varepsilon-\overline{u})]\, dx$$

$$= \int_{\partial\Omega} a\frac{\overline{u}(u_\varepsilon+\varepsilon-\overline{u})}{u_\varepsilon+\varepsilon}\frac{\partial u_\varepsilon}{\partial n}\, dx - \int_{\partial\Omega} a(u_\varepsilon+ \varepsilon - \overline{u})\frac{\partial\overline{u}}{\partial n}\, dx -$$

$$- a\int_\Omega\left[\nabla\left(\frac{\overline{u}(u_\varepsilon+ \varepsilon - \overline{u})}{u_\varepsilon+\varepsilon}\right)\nabla u_\varepsilon - \nabla(u_\varepsilon+\varepsilon - \overline{u})\,\nabla\overline{u}\right] dx -$$

$$- bc\int_\Omega \overline{u}(v_\varepsilon - \overline{v})^2\, dx$$

$$= - a\varepsilon\int_{\partial\Omega} \frac{\partial\overline{u}}{\partial n}\, dx - \int_\Omega a\overline{u}^2\left(\nabla\log\frac{u_\varepsilon+\varepsilon}{\overline{u}}\right)^2 dx -$$

$$- bc\int_\Omega \overline{u}(v_\varepsilon - \overline{v})^2\, dx \qquad \text{for all } t\in[0,\infty). \tag{20}$$

By integration with respect to time we get

$$0 \le \Lambda_\varepsilon(t_2) + \int_{t_1}^{t_2}\left(\int_\Omega a\overline{u}^2\left(\nabla\log\frac{u_\varepsilon+\varepsilon}{\overline{u}}\right)^2 + bc\overline{u}(v_\varepsilon - \overline{v})^2\, dx + a\varepsilon\int_{\partial\Omega}\frac{\partial\overline{u}}{\partial n}\, dx\right) d\tau$$

$$= \Lambda_\varepsilon(t_1) \quad \text{for all } t_1,t_2 \text{ with } 0 \le t_1 < t_2 < \infty. \tag{21}$$

We pass to the limit $\varepsilon \to 0$.
The continuous dependence of the solution of (17) on ε implies

$$\lim_{\varepsilon\to 0}[\|u_\varepsilon - u\| + \|v_\varepsilon - v\|]_{C(\overline{\Omega}\times[0,T])} = 0 \quad \text{for all } T\in(0,\infty) \tag{22}$$

and hence Lemma 21 p.78 implies

$$\lim_{\varepsilon\to 0} \|u_\varepsilon - u\|_{C^1([t_1,t_2]\times\overline{\Omega})} = 0 \quad \text{for all } t_1,t_2\in(0,\infty). \tag{23}$$

Recall that by the strong maximum principle p.123, the solution $\overline{u} \gneq 0$, $\overline{u} \in C^2(\overline{\Omega})$ of system (3) satisfies even

$$\overline{u}(x) > 0 \quad \text{for all } x \in \overline{\Omega}\setminus\partial\Omega \quad \text{and} \quad \partial\overline{u}/\partial n > 0 \quad \text{for all } x \in \partial\Omega. \tag{24}$$

For all functions $w \in E = C_o(\bar{\Omega}) \cap C^1(\bar{\Omega})$ we define the norm

$$\|w\|_{C_{\bar{u}}(\Omega)} = \sup_{x\in\Omega} |w(x)|/\bar{u}(x).$$

By (24) there exists a constant $K(\Omega,\bar{u})$ such that

$$\|w\|_{C_{\bar{u}}(\Omega)} \leq K(\Omega,\bar{u})\|w\|_{C^1(\bar{\Omega})} \quad \text{for all } w \in C_o(\bar{\Omega}) \cap C^1(\bar{\Omega}). \tag{25}$$

Hence (22)(23) imply

$$\lim_{\varepsilon\to 0} \|(u_\varepsilon - u)(.,t)\|_{C_{\bar{u}}(\Omega)} = 0 \quad \text{for all } t\in(0,\infty). \tag{26}$$

Furtheron the maximum principle implies that the solution (u,v) of system (1) satisfies

$$u(x,t) > 0 \qquad \text{for all } x \in \bar{\Omega}\setminus\partial\Omega,\ t\in(0,\infty); \tag{27}$$

$$\partial u/\partial \mathbf{n}(x,t) > 0 \qquad \text{for all } x \in \partial\Omega,\ t\in(0,\infty).$$

Hence there exists a continuous function $\alpha\colon t\in(0,\infty) \to \alpha(t)\in(0,1)$ such that

$$u(x,t) \geq 2\alpha(t)\bar{u}(x) \quad \text{for all } x \in \bar{\Omega},\ t\in(0,\infty).$$

Hence by (26) there exists $\varepsilon_o\in(0,1)$ such that

$$u_\varepsilon(x,t) \geq \alpha(t)\bar{u}(x) \quad \text{for all } x \in \bar{\Omega},\ t\in(0,\infty),\ \varepsilon\in[0,\varepsilon_o]. \tag{28}$$

Now we show that

$$\lim_{\varepsilon\to 0} \Lambda_\varepsilon(t) = \Lambda(t) \qquad \text{for all } t\in[0,\infty). \tag{29}$$

Consider at first the case that $t > 0$.

Then (22)(28) show that the integrant in the formula (19) defining is bounded uniformly for all $(x,\varepsilon) \in \bar{\Omega}\times[0,\varepsilon_o]$. Hence (22) implies (29).

Now consider the case $t = 0$. Since by assumption (7)

$$0 \leq \bar{u}[u_o+\varepsilon-\bar{u}-\bar{u}\log((u_o+\varepsilon)/\bar{u})] \leq \bar{u}[u_o+1-\bar{u}-\bar{u}\log(u_o/\bar{u})] \in L_1(\Omega),$$

the integrant in formula (19) defining $\Lambda_\varepsilon(0)$ can be estimated by an L_1-function uniformly for all $\varepsilon\in[0,1)$. Hence (22) and Lebesgues's dominanted convergence Theorem (Segal and Kunze [61] p. 72) imply (29) for $t = 0$.

Hence (29) is verified and we have shown that $\Lambda(t) < \infty$ for all $t\in[0,\infty)$.

Now we can pass to the limit $\varepsilon \to 0$ in formula (21). If $t_1 > 0$, (22)(29) yield the assertion (16) even with equality sign. If $t_1 = 0$, we get only an inequality. To see this, we restict the integration in (21) to any closed subset $[t,t_2]\times\bar\Omega_1 \subset (0,t_2]\times\Omega$. In the resulting inequality we pass to the limit $\varepsilon \to 0$. Hence (22)(23)(29) and the Lemma of Fatou (Segal and Kunze [61] p.69) imply for arbitrary $t_2 \in (0,\infty)$ that

$$\Lambda(t_2) + \int_t^{t_2}\int_{\Omega_1} a\bar u^2[\nabla \log(u(x,\tau)/\bar u]^2 + bc\bar u[v(x,\tau)-\bar v]^2\, dxd\tau$$

$$\leq \Lambda(0) \quad \text{for all } \bar\Omega_1 \subset \Omega,\ t\in(0,t_2).$$

Hence we get the assertion (16) with $t_1 = 0$. Thus the Lemma is proved.

Lemma 2

Let $p_0 \in [1,\infty]$ be arbitrary and take regular convenient initial data.

Then there exists a constant M as specified p.175 such that the solution (u,v) of system (1) satisfies the estimate

$$\sup_{0<t<\infty} m(t)^{\delta}\|u(t)\|_p \leq M \quad \text{for all } p \in [p_0,\infty]\cap(1,\infty], \quad \delta = n/p_0 - n/p. \tag{30}$$

Proof

Define the function $w \in C_0^2(\bar\Omega)$ by setting

$$w = \begin{cases} \varphi_0 & \text{if } \lambda \leq \lambda_0, \\ \bar u & \text{if } \lambda > \lambda_0. \end{cases}$$

In case $\lambda \leq \lambda_0$, (13)(15) imply the primary a priori estimate

$$⨍_\Omega w(x)[u + v^2](x,t)\,dx \leq M_1 \quad \text{for all } t\in[0,\infty). \tag{31}$$

We show that (31) holds in the case $\lambda > \lambda_0$, too.

Let the function $l\colon u\in(0,\infty) \to l(u)\in[0,\infty)$ be given by

$$l(u) = u - 1 - \log u.$$

Since $\quad l''(u) > 0 \quad$ for all $u\in(0,\infty)$,

Jensen's inequality implies

$$l\left[\frac{⨍_\Omega \bar u^2(x)[u(x,t)/\bar u(x)]\,dx}{⨍_\Omega \bar u^2(x)\,dx}\right] \leq \frac{⨍_\Omega \bar u^2(x)\, l[u(x,t)/\bar u(x)]\,dx}{⨍_\Omega \bar u^2(x)\,dx} \quad \text{for all } t\in(0,\infty). \tag{32}$$

$$\leq \Lambda(0)\ (a⨍_\Omega \bar u^2\,dx)^{-1} \leq M$$

Hence (31) follows.

We want to get rid of the weight function w in (31). Tor this end, choose $r\in(0,1/2)$ arbitrary and $s\in(0,1)$ such that $1 + 1/s = 1/r$.

Since the maximum principle implies

$$\partial w/\partial n(x,t) > 0 \quad \text{for all } x \in \partial\Omega,$$

we get

$$\int_\Omega w^{-s}(x)dx = M_2 < \infty. \tag{33}$$

Now Hölder's inequality and (31)(33) imply

$$\left[\int_\Omega u^r(x,t)\,dx\right]^{1/r} \leq \left[\int_\Omega w(x)u(x,t)\,dx\right]\left[\int_\Omega w^{-s}(x)\,dx\right]^{1/s} \tag{34}$$

$$\leq M_3 \quad \text{for all } t\in(0,\infty).$$

Applying the same reasoning to the component v, we get

$$\|u(t)\|_r + \|v(t)\|_{2r} \leq M_4 \quad \text{for all } t\in(0,\infty). \tag{35}$$

To the differential inequality

$$0 \leq u \quad \text{and} \quad u_t - \Delta u \leq \lambda u \quad \text{for all } x \in \Omega,\ t\in(0,\infty);$$

we apply Lemma 19 and 20 p.69 and 72. The weight function is $c \equiv \lambda$ and the primary a priori estimate is given by (35). We use the one-sided version. Thus we have $r_1 = r, \delta$ from above, $q_1 = q_2 = r_2 = \infty$, $\gamma = 1$. For $p \in [p_o,\infty]\cap(1,\infty], p - p_o$ small, there exist $\beta,\varepsilon\in(0,1)$ such that

$$(1-\beta)n/r < 1 + (1-\beta)n/p - \varepsilon, \quad (1-\beta)/r + \beta/p \leq 1\ , \quad \beta\delta < 1 - \varepsilon.$$

Hence assumptions (207)-(211) of Lemma 19 p.69 are satisfied. Hence assertion (214) of Lemma 19 proves (30) for $p - p_o$ small.
Next consider the case $p = \infty$, $\overline{\delta} = n/p_o$. We apply Lemma 20 p.72. Assumptions (231)-(233) are satisfied for $\beta\in(0,1)$ close to 1. Hence assertion (234) of Lemma 20 proves (30) for $p = \infty$. The general case follows by the interpolation inequality (124) p.51.
Thus the Lemma is proved.

In the following proofs we use the mild formulation of system (1a)-(1d):

$$u(t) - e^{\lambda t}S(t)u_o = -b\int_0^t e^{\lambda(t-s)}S(t-s)[uv](s)\,ds, \tag{36a}$$

$$v(t) - e^{-ct}v_o = a\int_0^t e^{-c(t-s)}u(s)\,ds \quad \text{for all } t\in(0,\infty). \tag{36b}$$

Lemma 3

Let $p_o, q_o, p_1, q \in [1,\infty]$, $\varepsilon, \hat{\varepsilon} \in (0,1)$ satisfy

$$n/p_o - n/q + \varepsilon \leq 1 \ , \tag{37}$$

$$n/p_o + n/q_o < n/p_1 + 1 - \hat{\varepsilon} \quad \text{and} \quad n/p_o < n/(2p_1) + 1 - \hat{\varepsilon}/2 \ . \tag{38}$$

Take regular convenient initial data.

Then there exists a constant M_5 as specified p.175 such that the solution (u,v) of the system (1) satisfies

$$\|v(t) - e^{-ct}v_o\|_q \leq m(t)^{\varepsilon} M_5 \quad \text{for all } t \in (0,\infty), \tag{39}$$

$$\|u(t) - e^{\lambda t}S(t)u_o\|_{p_1} \leq t^{\hat{\varepsilon}} M_5 \quad \text{for all } t \in (0,1). \tag{40}$$

Proof

There exists $p \in [q,\infty] \cap [p_o,\infty] \cap (1,\infty]$ such that $1 - \varepsilon \leq \delta = n/p_o - n/p < 1$. Equation (36b) and estimate (30) imply

$$\begin{aligned} \|v(t) - e^{-ct}v_o\|_q &\leq a\,\|u\|_{p,\delta,t} \int_0^t e^{-c(t-\tau)} m(\tau)^{-\delta}\, d\tau \\ &\leq m(t)^{\varepsilon} M_6 \quad \text{for all } t \in (0,\infty). \end{aligned}$$

Thus (39) is proved. Next we prove (40).
There exist $s, p, q \in [1,\infty]$, $\delta \in [0,1)$ such that

$$1/p + 1/q \leq 1/s \ , \tag{41}$$

$$n/s - n/p_1 + \delta + \hat{\varepsilon} < 1 \ , \tag{42}$$

$$n/p_o - n/p < \delta \ , \quad n/p_o - n/q < 1 \ , \quad q \leq q_o \ . \tag{43}$$

By (36a) and Hölder's inequality we estimate

$$\|u(t) - e^{\lambda t}S(t)u_o\|_{p_1} \leq b\|u\|_{p,\delta,t}\|v\|_{q,o,t} \int_0^t e^{\lambda(t-\tau)} \|S(t-\tau)\|_{p_1,s} m(\tau)^{-\delta} d\tau \quad \text{for all } t \in (0,\infty). \tag{44}$$

Via assumption (42) and Lemma 6 p.31, the integral in (44) can be estimated by $M_7 t^{\hat{\varepsilon}}$ for all $t \in (0,1)$.
Via assumption (43) and (30) of Lemma 2, $\|u\|_{p,\delta,t}$ can be estimated by M_8.
Via assumption (43) and (39) above, $\|v\|_{q,o,t}$ can be estimated by M_9. Thus

$$\|u(t) - e^{\lambda t}S(t)u_o\|_{p_1} \leq bM_8M_9M_7t^{\hat{\varepsilon}} \qquad \text{for all } t \in (0,1),$$

which proves assertion (40).
Thus the Lemma is proved.

Lemma 4

Let $p_0, q_0, p_1 \in [1,\infty]$, $\delta \in [0,1)$ satisfy

$$n/p_0 < 1 + \delta \ , \quad n/p_0 + n/p_1 < n + \delta \ , \quad n/p_0 - n/p_1 < \delta \ ,$$
$$n/q_0 < 1 \ , \quad 1/q_0 \leq 1/p_1 \ , \quad 1/q_0 + 1/p_1 \leq 1 \ . \tag{45}$$

For i = 1,2 let (u_i, v_i) be the solutions of system (1) for regular convenient initial data (u_{0i}, v_{0i}).

Then there exists a constant M_{10} as specified p.175 such that

$$\|u_1 - u_2\|_{p_1,\delta,t} + \|v_1 - v_2\|_{p_1,0,t} \leq M_{10}[\|u_{01}-u_{02}\|_{p_0} + \|v_{01}-v_{02}\|_{q_0}] \tag{46}$$

for all $t \in (0, 1/M_{10})$.

Proof

There exist $s, q \in [1,\infty]$, $\varepsilon \in (0,1)$ such that

$$0 \leq n/p_0 - n/p_1 < \delta < 1 \ , \tag{47}$$

$$n/s - n/p_1 < 1 - \varepsilon \ , \quad 1/p_1 + 1/q \leq 1/s \ , \tag{48}$$

$$n/p_0 - n/q < \delta < 1 \ , \quad q \leq q_0 \ . \tag{49}$$

Subtracting the integral equation (36a) for i = 1,2 we estimate

$$m(t)^{\delta}\|(u_1-u_2)(t)\|_{p_1} \leq m(t)^{\delta}\|S(t)\|_{p_1,p_0}\|u_{01}-u_{02}\|_{p_0} + \tag{50}$$

$$+ m(t)^{\delta}\int_0^t \|S(t-\tau)\|_{p_1,s} m(\tau)^{-\delta} d\tau \Big[\|u_1-u_2\|_{p_1,\delta,t}\|v_1\|_{q,0,t} + \|u_2\|_{q,\delta,t}\|v_1-v_2\|_{p_1,0,t}\Big]$$

$$\leq K_1\|u_{01}-u_{02}\|_{p_0} + m(t)^{\varepsilon}M_{11}[\|u_1-u_2\|_{p_1,\delta,t} + \|v_1-v_2\|_{p_1,0,t}] \tag{51}$$

for all $t \in (0,\infty)$.

The first part (50) of this estimate uses Hölder's inequality and (48). We explain the second part (51). The first summand is estimated by (47) and assertion (30) of Lemma 3(i) p.25. In the second summand, the integral can be estimated via Lemma 6 p.31 and (48) by $M_{12}m(t)^{\varepsilon}$. In the large bracket, $\|u_2\|_{q,\delta,t}$ is estimated by (49) and Lemma 2, whereas $\|v_1\|_{q,0,t}$ is estimated by (49) and assertion (39) of Lemma 3.

We turn to the integral equation (36b) for the component v and estimate

$$\|(v_1-v_2)(t)\|_{p_1} \leq e^{-ct}\|v_{01}-v_{02}\|_{q_0} + M_{13}m(t)^{1-\delta}\|u_1-u_2\|_{p_1,\delta,t} \tag{52}$$

for all $t \in (0,\infty)$.

Let $\rho = \min\{\varepsilon, 1-\delta\}$ and define the function $\varphi: t\in(0,\infty) \to \varphi(t)\in[0,\infty)$ by

$$\varphi(t) = \|u_1-u_2\|_{p_1,\delta,t} + \|v_1-v_2\|_{p_1,o,t}.$$

By estimates (51)(52) we get

$$\varphi(t) \leq K_2[\|u_{o1}-u_{o2}\|_{p_o} + \|v_{o1}-v_{o2}\|_{q_o}] + m(t)^{\rho} M_{14}\varphi(t) \quad \text{for all } t\in(0,\infty). \tag{53}$$

Now choose $h\in(0,1)$ small enough such that

$$m(h)^{\rho} M_{14} \leq 1/2.$$

Then (53) implies

$$\varphi(t) \leq 2K_2[\|u_{o1}-u_{o2}\|_{p_o} + \|v_{o1}-v_{o2}\|_{q_o}] \quad \text{for all } t\in(0,h),$$

which proves the assertion (46). Thus the Lemma is proved.

Lemma 5

Let $p_o\in[1,\infty]$, $q_o,q\in[2,\infty]$ satisfy the hypotheses of the Theorem p.173. Take regular convenient initial data.

Then there exists $\varepsilon\in(0,1)$ and a constant M_{15} as specified p.175 such that the solution (u,v) of system (1) satisfies

$$\|u(t) - S(t)u_o\|_{p_o} \leq m(t)^{\varepsilon}M_{15} \quad \text{for all } t\in(0,\infty); \tag{54}$$

$$\|v(t) - v_o\|_q \leq m(t)^{\varepsilon}M_{15} \quad \text{for all } t\in(0,\infty); \tag{55}$$

$$\sup_{0<t<\infty} m(t)^{\delta}\|u(t)\|_p \leq M_{15} \quad \text{for all } p\in[p_o,\infty],\ \delta = n/p_o - n/p; \tag{56}$$

$$\int_0^1 \|(u + uv)(\tau)\|_1 d\tau \leq M_{15}. \tag{57}$$

For $i = 1,2$, let (u_i,v_i) be the solutions of system (1) for regular convenient initial data (u_{oi},v_{oi}). Then the solutions depend continuously on the initial data in the following sense:

$$\|u_1-u_2\|_{q,1-\varepsilon,t} + \|v_1-v_2\|_{q,o,t} \leq M_{15}e^{M_{15}t}[\|u_{o1}-u_{o2}\|_{p_o} + \|v_{o1}-v_{o2}\|_{q_o}] \quad \text{for all } t\in(0,\infty). \tag{58}$$

Proof

Apply Lemma 3 estimate (40) with $p_1 := p_o$. Then assertion (54) follows for t restricted by $t\in(0,1)$. Now estimate (30) of Lemma 2 proves (54) and (56). Assertion (55) was already shown in Lemma 3. Finally assertion (55)(56) with $p = q = 2$, $\delta = n/p_o - n/2 < 1$ imply (57).

It remains to prove (58).

Applying Lemma 4 with $p_1 := q$ proves assertion (58) with the restriction $t \in (0, t_o]$ for some $t_o = 1/M_{10}$.

Now we apply Lemma 4 with $p_o := q_o := p_1 := q$ and initial data $(u,v)(t_o + ih)$ successively for $i = 0,1,2 \ldots$. Since by (55)(56)

$$\sup_{i \in N} \|(u+v)(t_o+ih)\|_q \leq M_{16},$$

the time step $h \in (0,\infty)$ can be chosen such that assertion (46) of Lemma 4 yields $h = 1/M_{18}$ and

$$\sup\{\|(u_1-u_2)(t)\|_q + \|(v_1-v_2)(t)\|_q \mid t \in [t_o+ih, t_o+ih+h]\} \leq \tag{59}$$

$$\leq M_{17}[\|(u_1-u_2)(t_o+ih)\|_q + \|(v_1-v_2)(t_o+ih)\|_q] \quad \text{for all } i \in \{0\} \cup N.$$

Now assertion (58) follows by induction on i.

Thus the Lemma is proved.

We return to the proof of the Theorem.

If $q < \infty$, one may assume $q_o < \infty$, too. Let arbitrary convenient initial data $(u_o, v_o) \in \mathcal{K}$ as assumed in the Theorem be given. Then there exists a sequence (u_{om}, v_{om}) of regular convenient data (satisfying (11) p.174) such that

$$\lim_{m\to\infty} [\|u_{om} - u_o\|_{p_o} + \|v_{om} - v_o\|_{q_o}] = 0; \tag{60}$$

$$\|u_{om}\|_{p_o} \leq \|u_o\|_{p_o}, \quad \|v_{om}\|_{q_o} \leq \|v_o\|_{q_o} \quad \text{for all } m \in N; \tag{61}$$

$$\Lambda(u_{om}, v_{om}) \leq \Lambda(u_o, v_o) \quad \text{for all } m \in N \tag{62}$$

with the Lyapunov functional given by (13)(14).

Let (u_m, v_m) be the global solution of system (1) for initial data (u_{om}, v_{om}). The constants M occuring in Lemma 2 through 5 can be chosen independent of $m \in N$ by (61)(62). Hence assertion (58) of Lemma 5 implies that the sequence (u_m, v_m) converges to $(u,v) \in E_{p_o,0,\infty} \times E_{q,0,\infty}$ in the sense

$$\lim_{m\to\infty} [\|u_m - u\|_{q,1-\varepsilon,t} + \|v_m - v\|_{q,0,t}] = 0 \quad \text{for all } t \in (0,\infty) \tag{63}$$

and estimates (54)(55)(56)(57) of Lemma 5 hold for the limit (u,v). It is straightforward to derive from (36a)(36b) that (u,v) is indeed the mild solution of system (1) for initial data (u_o, v_o) (see p.122).

We are ready to consider the asymptotic behavior. Let $w \in C^2(\overline{\Omega})$ be as given on p.179.

<u>Lemma 6</u>

$$\lim_{t\to\infty} \int_\Omega w(x)[v(x,t) - \overline{v}(x)]^2\, dx = 0. \tag{64}$$

<u>Proof</u>

Define the function $V: t\in[0,\infty) \to V(t)\in[0,\infty)$ by setting

$$V(t) = \int_\Omega w(x)[v(x,t) - \overline{v}(x)]^2\, dx.$$

The estimates of the Lyapunov functional in Lemma 1 imply

$$V(t) + c\int_0^1 V(\tau)d\tau \leq M_{19} \quad \text{for all } t\in[1,\infty). \tag{65}$$

In the case of classical solutions (e.g. regular initial data) explicit differentiation of $V(t)$ yields

$$\frac{dV}{dt} = \int_\Omega 2w(v-\overline{v})(-cv+au)\,dx = -\int_\Omega 2cw(v-\overline{v})^2 dx + \int_\Omega 2aw(v-\overline{v})(u-\overline{u})\,dx$$

$$\leq 2(a^2/c)\int_\Omega w(u-\overline{u})^2 dx.$$

Hence estimate (30) of Lemma 2 implies

$$\frac{dV}{dt} \leq M_{20} \qquad \text{for all } t\in[1,\infty). \tag{66}$$

Integration with respect to time yields

$$V(t_2) - V(t_1) \leq M_{20}(t_2 - t_1) \quad \text{for all } t_1,t_2\in[1,\infty),\ t_1 < t_2. \tag{67}$$

This estimate holds for arbitrary convenient initial data, too, by the approximation process p.184.

Now we argue by contradiction: Assume that the assertion (64) is false. Then there exists a sequence (t_m) and $\theta > 0$ such that

$$\lim_{m\to\infty} t_m = \infty \quad \text{and} \quad V(t_m) > \theta \quad \text{for all } m \in N.$$

Hence (67) implies $\int_0^\infty V(\tau)d\tau = \infty$ contradicting (65). Thus the Lemma is proved.

<u>Lemma 7</u>

$$\lim_{t\to\infty} \|u(t) - \overline{u}\|_{C^\nu(\overline{\Omega})} = 0 \tag{10a}$$

$$\lim_{t\to\infty} \|v(t) - \overline{v}\|_q = 0 \tag{10b}$$

Proof

In the right-hand side of the differential equation (1a), we consider u and v as known functions of (x,t). Thus define the function

$$F: (x,t)\in\overline{\Omega}\times(0,\infty) \to F(x,t)\in R$$

by

$$F(x,t) = [u(\lambda - bv)](x,t).$$

We use the following mild formulation of (1a)(1c):

$$u(t_0+t) = S(t)u(t_o) + \int_0^t S(t-s)F(t_0+s)\,ds \quad \text{for all } t_0,t\in(0,\infty). \tag{68}$$

Estimates (30)(39) imply

$$\sup_{1\le s<\infty} \|F(s)\|_q \le M_{21}. \tag{69}$$

We construct a limit set.

Choose $\nu\in(0,2)$, $\mu\in(0,1)$, $T\in(0,\infty)$ such that $n/q + \nu/2 + \mu < 1$.

Define the space of Hölder continuous functions

$$X^{\mu,\nu,T} = C^{\mu}([0,T],C^{\nu}(\Omega))$$

and the following limit set for the component u:

$$\omega = \{U \in X^{\mu,\nu,T} \mid \text{there exists a sequence } t_m \to \infty \text{ such that } \lim_{m\to\infty} \|u(.,t_m+.) - U\|_{X^{\mu,\nu,T}} = 0.\}$$

By (68)(69) Lemma 21 p.78 implies

$$\sup_{2\le s<\infty} \|u(.,s+.)\|_{X^{\hat{\mu},\hat{\nu},T}} \le M_{22} \quad \text{for } \hat{\mu}\in(\mu,1),\ \hat{\nu}\in(\nu,2) \text{ with } n/q+\hat{\nu}/2+\hat{\mu} < 1.$$

For $\mu < \hat{\mu}$, $\nu < \hat{\nu}$ the embedding $X^{\hat{\mu},\hat{\nu},T} \subset X^{\mu,\nu,T}$ is compact. Hence it is easy to see that the limit set ω is nonvoid and compact.

We show that ω consist of a single point, namely the equilibrium $\overline{u}$ (considered as a time-independent function). Take an arbitrary point $U \in \omega$. By definition of the limit set ω there exists a sequence (t_m) with

$$\lim_{m\to\infty} t_m = \infty \quad\text{and}\quad \lim_{m\to\infty} \|u(.,t_m+.) - U\|_{X^{\mu,\nu,T}} = 0. \tag{70}$$

The differential equation (1b) for the component v implies

$$v(x,t_m+t) - v(x,t_m) = \int_0^t (-cv + au)(x,t_m+s)\,ds \quad \text{for all } x \in \overline{\Omega},\ t\in(0,\infty),\ m \in N. \tag{71}$$

Introduce the norm $\|v\|_w^2 = \int_\Omega w(x)v^2(x)\,dx$ for all $v \in L_2(\Omega)$.

By (64) of Lemma 6 and (70) both sides of equation (71) converge for $m \to \infty$ in the norm $\| \ \|_W$. Hence we get in the limit

$$0 = \int_0^t [-c\bar{v}(x) + aU(x,s)]ds \quad \text{for all } x \in \bar{\Omega},\ t \in [0,T]$$

and

$$U(x,s) = (c/a)\bar{v}(x) = \bar{u}(x) \quad \text{for all } x \in \bar{\Omega},\ t \in [0,T].$$

Thus the limit set ω consists indeed of the single point $\bar{u}$. Hence

$$\lim_{t\to\infty} \|u(.,t+.) - \bar{u}\|_{X^{\mu,\nu,T}} = 0,$$

which proves assertion (10a).

It remains to show (10b). The differential equation (1b) implies

$$v(t_2) = e^{-c(t_2-t_1)}v(t_1) + a\int_0^{t_2-t_1} e^{-c(t_2-t_1-s)}u(t_1+s)\,ds$$

and

$$[v(t_2) - \bar{v}] = e^{-c(t_2-t_1)}[v(t_1) - \bar{v}] +$$

$$+ a\int_0^{t_2-t_1} e^{-c(t_2-t_1-s)}[u(t_1+s) - \bar{u}]\,ds$$

for all $t_1, t_2 \in (0,\infty)$ with $t_1 < t_2$.

Using (39) we can estimate

$$\|v(t_2) - \bar{v}\|_q \le e^{-c(t_2-t_1)} M + (a/c)\sup\{\|u(s) - \bar{u}\|_q \mid s \in [t_1,\infty]\}.$$

By (10a) the right-hand side tends to zero for $t_1 \to \infty$, $t_2 - t_1 \to \infty$. Hence (10b) follows. Thus the Lemma and the whole Theorem is proved.

The Volterra-Lotka Model

In the following we study the classical Volterra-Lotka model. This is a predator-prey model which assumes the mass action law for the interaction of the two species, predator and prey. Denote the density of the prey by u and of the predator by v. Since we are concerned with a spatially distributed system these are functions of the space variable $x \in \Omega \subset R^N$ and of time t. We assume that the domain Ω is bounded. The dynamics are governed by the following reaction-diffusion system:

$$u_t - D_u \Delta u = u(f - bv) \qquad (1a)$$

$$\text{for all } x \in \Omega,\ t > 0.$$

$$v_t - D_v \Delta v = v(-g + au) \qquad (1b)$$

In general a,b,f,g are assumed to be positive. An inhomogeneous environment can be described by taking for a,b,f,g functions of the space variable x. Symbiosis and saturation can be included by letting f and g depend on u and v, too. Of course, for these generalized systems, the asymptotic behavior is not known. To give an example, Mimura and Murray [40] have studied the autonomous system with nonmonotone functions $f = f(u)$ and $g = g(v)$ in order to explain the formation of spatial patterns in the planktonic system. Some special cases can be studied rather completely. In [57] and a forthcoming paper the author considers the case that $f = f(x,u)$ in decreasing in u and $g = g(x,v)$ is increasing in v and only one component diffuses. A related competition system with one diffusing and one nondiffusing component is studied by Mottoni et.al.[44].

In the following we restrict ourselves to some simple cases. Assume that $D_u, D_v \in [0,\infty)$ and $a,b,f,g \in (0,\infty)$ are constants. Then system (1a)(1b) has a unique positive equilibrium $(\bar{u},\bar{v}) \in (0,\infty)^2$ given by

$$\bar{u} = g/a \quad \text{and} \quad \bar{v} = f/b. \qquad (2)$$

We study system (1a)(1b) in the case of Dirichlet boundary conditions (1c) or Neumann boundary conditions (1d) and initial conditions (1e).

$$u(x,t) = \bar{u} \quad \text{and} \quad v(x,t) = \bar{v} \quad \text{for all } x \in \partial\Omega,\ t > 0 \qquad (1c)$$

or

$$\partial u/\partial n(x,t) = 0 \quad \text{and} \quad \partial v/\partial n(x,t) = 0 \quad \text{for all } x \in \partial\Omega,\ t > 0\ ; \qquad (1d)$$

finally

$$u(x,0) = u_o(x) \quad \text{and} \quad v(x,0) = v_o(x) \quad \text{for all } x \in \Omega. \qquad (1e)$$

<u>Theorem 1</u> (Boundedness and convergence to equilibrium)

Let space dimension N be arbitrary, $n = N/2$.
Let $p_o, q_o \in [1,\infty]$ satisfy $n/p_o < 1$.
Take initial data $u_o \in L_{p_o}(\Omega)$, $v_o \in L_{q_o}(\Omega)$ such that $u_o \geqq 0$, $v_o \geqq 0$ and $|\log u_o| + |\log v_o| \in L_1(\Omega)$.
Assume for both diffusion coefficients $D_u > 0$ and $D_v > 0$.

Then the Volterra-Lotka system (1a)(1b) with Dirichlet boundary condition (1c) or Neumann boundary condition (1d) has a global classical solution (u,v) on the time interval $(0,\infty)$ satisfying

$$\lim_{t\to 0} \|u(t) - u_o\|_1 = 0; \tag{3a}$$

$$\lim_{t\to 0} \|v(t) - v_o\|_{q_o} = 0 \quad \text{if } q_o < \infty; \tag{3b}$$

$$\sup_{0<t<\infty} m(t)^{\delta} \|u(t)\|_p < \infty \quad \text{for all } p \in [p_o,\infty],\ \delta = n/p_o - n/p; \tag{4a}$$

$$\sup_{0<t<\infty} m(t)^{\eta} \|v(t)\|_q < \infty \quad \text{for all } q \in [q_o,\infty],\ \eta = n/q_o - n/q. \tag{4b}$$

In the case of Dirichlet boundary conditions (1c), the solution approaches the equilibrium $(\bar{u},\bar{v})$ for $t \to \infty$ in the following sense:
There exist constants $M, \lambda \in (0,\infty)$ such that

$$\|u(t) - \bar{u}\|_{C^2(\bar{\Omega})} + \|v(t) - \bar{v}\|_{C^2(\bar{\Omega})} \leqq M\, e^{-\lambda t} \quad \text{for all } t \in [1,\infty). \tag{5}$$

<u>Remark</u>
The asymptotic behavior in the case of Neumann boundary conditions (1d) is more involved and will not be investigated here (see Alikakos [1] and Rothe [57] for this case).

By some modifications, the degenerate case that only one species diffuses can be included. We consider two cases: diffusing prey and sedentary predator corresponding to the upper signs as well as sedentary prey and diffusing predator corresponding to the lower signs:

$$u_t - D\Delta u = u(\pm f \mp bv) \tag{6a}$$

for all $x \in \Omega$, $t > 0$.

$$v_t = v(\mp g \pm au) \tag{6b}$$

We assume the Dirichlet boundary condition (6c) and initial conditions (6d):

$$u(x,t) = \bar{u} \qquad \text{for all } x \in \partial\Omega,\ t > 0; \tag{6c}$$

$$u(x,0) = u_o(x) \quad \text{and} \quad v(x,0) = v_o(x) \quad \text{for all } x \in \Omega. \tag{6d}$$

Theorem 2 (Degenerate cases with one nondiffusing species)

Take arbitrary space dimension N, n = N/2.
Let $p_o, q_o \in [1,\infty)$ satisfy

$$n/p_o < 1 , \quad n/q_o < 1 . \tag{7}$$

Take initial data $u_o \in L_{p_o}(\Omega)$, $v_o \in L_{q_o}(\Omega)$ such that
$u_o \geqq 0$, $v_o \geqq 0$ and $|\log u_o| + |\log v_o| \in L_1(\Omega)$.

Then the Volterra-Lotka system (6a)-(6d) has a global L_1-mild solution on the time interval $[0,\infty)$ satisfying

$$\lim_{t\to 0} \|u(t) - u_o\|_{p_o} = 0; \tag{8a}$$

$$\lim_{t\to 0} \|v(t) - v_o\|_{q_o} = 0; \tag{8b}$$

$$\sup_{0<t<\infty} \|(u + v + |\log u| + |\log v|)(t)\|_1 < \infty; \tag{9}$$

$$\int_1^\infty \int_\Omega [u(x,t) - \bar{u} - \bar{u}\log(u(x,t)/\bar{u})]^2 \, dxdt < \infty; \tag{10}$$

$$\int_1^\infty \int_\Omega (\nabla \log u(x,t))^2 \, dxdt < \infty. \tag{11}$$

More information can be given in the following two cases:

(a) space dimension N = 1 and lower signs (diffusing predator),
(b) arbitrary space dimension and upper signs (diffusing prey).

If (a) or (b) hold, then

$$\sup_{0<t<\infty} m(t)^\delta \|u(t)\|_p < \infty \quad \text{for all } p \in [p_o,\infty],\ \delta = n/p_o - n/p. \tag{12}$$

Take positive initial data $u_o, v_o \in C^{2+\alpha}(\bar{\Omega})$ satisfying (44) p.198. Then

$$\lim_{t\to\infty} \|u(t) - \bar{u}\|_1 = 0; \tag{13a}$$

$$\lim_{t\to\infty} \int_\Omega \varphi(x)[v(x,t) - \bar{v}] \, dx = 0 \quad \text{for all test functions } \varphi \in C_o(\bar{\Omega}); \tag{13b}$$

$$\lim_{t\to\infty} \int_\Omega [u(x,t) - \bar{u} - \bar{u}\log(u(x,t)/\bar{u})] \, dx = 0; \tag{14a}$$

$$\lim_{t\to\infty} \int_\Omega \varphi(x)[v(x,t) - \bar{v} - \bar{v}\log(v(x,t)/\bar{v})]\,dx \tag{14b}$$

exists for all test functions $\varphi \in C(\bar{\Omega})$.

In case of space dimension N = 1, we get even

$$\lim_{t\to\infty} \|u(t) - \bar{u}\|_{C^\nu(\bar{\Omega})} = 0 \quad \text{for all } \nu\in(0,1); \tag{15}$$

$$\lim_{t\to\infty} \int_\Omega [\nabla u(x,t)]^2\,dx = 0. \tag{16}$$

Proof of Theorem 1

The set of convenient initial data consists of all $(u_o,v_o) \in L_{p_o}(\Omega) \times L_{q_o}(\Omega)$ as specified in the Theorem.

The set $\mathcal{R}$ of regular initial data consists of all $(u_o,v_o) \in C^{2+\alpha}(\bar{\Omega}) \times C^{2+\alpha}(\bar{\Omega})$ such that $u_o(x) > 0$ and $v_o(x) > 0$ for all $x \in \bar{\Omega}$ and the following compatibility conditions hold on the boundary:

$$u_o(x) = \bar{u},\ \Delta u_o(x) = 0 \quad \text{and} \quad v_o(x) = \bar{v},\ \Delta v_o(x) = 0 \quad \text{for all } x \in \partial\Omega \tag{17}$$

By Theorem 1 p.111 the system (1a)(1b) with Dirichlet boundary condition (1c) or Neumann boundary condition (1d) and initial condition (1e) has a unique classical solution (u,v) for initial data $(u_o,v_o) \in \mathcal{K} \cap \mathcal{R}$ on some maximal time interval $[0,T_{max})$. By the Comparison Theorem p.123

$$u(x,t) > 0 \quad \text{and} \quad v(x,t) > 0 \quad \text{for all } x \in \bar{\Omega},\ t\in[0,T_{max}). \tag{18}$$

Furtheron, it is easy to show that the solution (u,v) grows at most exponentially in time (see e.g. Williams and Chow [69] for even more general ecological systems with a food pyramide structure). Hence the explosion property (8) p.112 implies $T_{max} = \infty$.

In the following Lemma 1,2 and 3 we restrict ourselves to regular convenient initial data.

Lemma 1 constructs Lyapunov functionals for the systems (1) and (6). It turns out to be of advantage to introduce a variable test function $\varphi \in C^2(\bar{\Omega})$ into these functionals.

Lemma 2 uses the methods of part I (Lemma 20 p.72) to derive uniform a priori estimates from the primary estimate given by the Lyapunov functional.

Lemma 3 estimates how the initial data are approximated for $t \to 0$.

Let $\varphi \in C^2(\bar{\Omega})$ be a variable test function. Define the function

$$L: (u,\bar{u}) \in (0,\infty) \times (0,\infty) \to L(u,\bar{u}) \in [0,\infty) \quad \text{by}$$

$$L(u,\bar{u}) = u - \bar{u} - \bar{u} \log(u/\bar{u}).$$

Let $(u,v) = (u,v)(x,t)$ be the solution of system (1). Define

$$l: (x,t) \in \bar{\Omega} \times [0,\infty) \to l(x,t) \in [0,\infty) \quad \text{by}$$

$$l(x,t) = aL(u(x,t),\bar{u}) + bL(v(x,t),\bar{v}).$$

Finally define the functional $\Lambda_\varphi: t \in [0,\infty) \to \Lambda_\varphi(t) \in R$ by setting

$$\Lambda_\varphi(t) = \int_\Omega \varphi(x) l(x,t) dx = \tag{19}$$

$$= \int_\Omega \varphi(x) [a[u(x,t) - \bar{u} - \bar{u}\log(u(x,t)/\bar{u})] + b[v(x,t) - \bar{v} - \bar{v}\log(v(x,t)/\bar{v})]] \, dx$$

The functional Λ_1 with test function $\varphi \equiv 1$ will be denoted by Λ.

In the following p. 192-206 we denote by the letter M constants depending only on the exponents, the domain Ω, the constants a,b,f,g and D_u, D_v, the norms $\|u_o\|_{p_o}$, $\|v_o\|_{q_o}$ and the initial value $\Lambda(0)$ of the functional.

<u>Lemma 1</u>

Take positive regular initial data.

(i) Consider system (1) with Dirichlet boundary conditions (1c) or the degenerate system (6).

For any test function $\varphi \in C^2(\bar{\Omega})$, the functional Λ_φ is well defined and

$$\begin{aligned} \Lambda_\varphi(t_1) &- \int_{t_1}^{t_2}\int_\Omega D_u(\Delta\varphi)L(u,\bar{u}) + D_v(\Delta\varphi)L(v,\bar{v}) \, dxdt + \\ &+ \int_{t_1}^{t_2}\int_\Omega \bar{u}D_u\varphi(\nabla\log u)^2 + \bar{v}D_v\varphi(\nabla\log v)^2 \, dxdt \\ &= \Lambda_\varphi(t_2) \quad \text{for all } t_1,t_2 \in [0,\infty) \text{ with } t_1 < t_2. \end{aligned} \tag{20}$$

If the test function satisfies $\varphi \geqq 0$ and $-\Delta\varphi \geqq 0$, then

$$\Lambda_\varphi(t) \geqq 0 \quad \text{and} \quad d\Lambda_\varphi/dt \leqq 0 \quad \text{for all } t \in [0,\infty). \quad \text{Hence}$$

$$\lim_{t\to\infty} \Lambda_\varphi(t) = \Lambda_\infty \in [0,\infty) \quad \text{exists;} \tag{21}$$

$$D_u\int_0^\infty\int_\Omega (-\Delta\varphi)L(u,\bar{u}) + \varphi(\nabla\log u)^2 \, dxdt \leqq M\Lambda_\varphi(0); \tag{22a}$$

$$D_v\int_0^\infty\int_\Omega (-\Delta\varphi)L(v,\bar{v}) + \varphi(\nabla\log v)^2 \, dxdt \leqq M\Lambda_\varphi(0). \tag{22b}$$

(ii) Consider system (1) with Neumann boundary conditions (1d).

Then formula (20)(21)(22a)(22b) hold only with the test function $\varphi \equiv 1$.

(iii) In both cases (i) and (ii)

$$\sup_{0<t<\infty} \|(u + v + |\log u| + |\log v|)(t)\|_1 \leq M. \tag{23}$$

The constants M in (22a)(22b)(23) depend only on a,b,f,g, the domain Ω and $\Lambda(0)$.

Proof

A straightforward calculation using the differential equations (1a)(1b) shows that

$$d\Lambda_\varphi/dt = \int_\Omega \varphi(1-\bar{u}/u)D_u\Delta u + \varphi(1-\bar{v}/v)D_v\Delta v \, dx \equiv I_u + I_v. \tag{24}$$

Both terms I_u and I_v can be integrated by parts:

$$I_u = -D_u\int_\Omega \nabla[\varphi(1-\bar{u}/u)]\, \nabla u \, dx + D_u\int_{\partial\Omega} \varphi(1-\bar{u}/u)\, \frac{\partial u}{\partial n}\, dx \tag{25}$$

$$= \bar{u}D_u\int_\Omega \varphi\, \nabla(1/u)\, \nabla u\, dx - D_u\int_\Omega \nabla\varphi\, \nabla L(u,\bar{u})\, dx$$

$$= -\bar{u}D_u\int_\Omega \varphi(\nabla\log u)^2\, dx + D_u\int_\Omega (\Delta\varphi)L(u,\bar{u})\, dx - \tag{26}$$

$$- D_u\int_{\partial\Omega} \frac{\partial\varphi}{\partial n}\, L(u,\bar{u})\, dx.$$

The boundary terms in (25)(26) vanish in case of Dirichlet boundary conditions, since $u = \bar{u}$ and $L(u,\bar{u}) = 0$ on $\partial\Omega$ as well as in case of Neumann boundary conditions with $\varphi \equiv 1$. Hence (20) follows. Now (21)(22a)(22b) are straightforward consequences.

It remains to show (23). Since the function

$L(.,\bar{u}): u\in(0,\infty) \to L(u,\bar{u})\in[0,\infty)$ is convex, by Jensen's inequality

$$L\left(\int_\Omega u(x,t)dx\, ,\, \bar{u}\right) \leq \int_\Omega L(u,\bar{u})dx \leq \Lambda(0)/(a|\Omega|) \quad \text{for all } t\in[0,\infty). \tag{27}$$

A similar estimate holds for the component v. Hence (23) follows. Thus the Lemma is proved.

Remark

The degenerate system can be included, but the result for the nondiffusing component is indeed weaker, since (22b) become void.

Lemma 2

Let $p_o, q_o, q_1, q_2 \in [1,\infty]$ satisfy

$$n/p_o < 1 \quad \text{and} \quad n/p_o < n/q_1 + 1/q_2 \,.$$

Take regular convenient initial data.

Then there exists a constant M as specified p.192 such that the solution (u,v) of the system (1) satisfies

$$\sup_{0<t<\infty} m(t)^{\delta} \|u(t)\|_p \leq M \quad \text{for all } p \in [p_o,\infty],\ \delta = n/p_o - n/p; \tag{28a}$$

$$\sup_{0<t<\infty} m(t)^{\eta} \|v(t)\|_q \leq M \quad \text{for all } q \in [q_o,\infty],\ \eta = n/q_o - n/q; \tag{28b}$$

$$\|u\|_{q_1,q_2} \leq M. \tag{29}$$

Proof

To be definite, we consider the case of Dirichlet boundary conditions. The case of Neumann boundary conditions is even simpler.

Denote by $S_u(t)$ and $S_v(t)$ the semigroups generated by $-D_u\Delta$ and $-D_v\Delta$ in the domain Ω with Dirichlet boundary conditions on $\partial\Omega$. Let

$$u_1(x,t) = \bar{u} - [S_u(t)\bar{u}](x) \tag{30a}$$

$$\text{for all } x \in \Omega,\ t \in [0,\infty)$$

$$v_1(x,t) = \bar{v} - [S_v(t)\bar{v}](x) \tag{30b}$$

be the solutions ot the homogeneous heat equation with Dirichlet boundary conditions (1c) and zero initial conditions. In the mild formulation of the initial-boundary value problem (1a)(1b)(1c)(1e) additional terms occur because of the inhomogeneous boundary condition:

$$u(t) = S_u(t)u_o + u_1(t) + \int_0^t S_u(t-s)[u(f-bv)](s)\,ds \tag{31a}$$

$$\text{for all } t \in (0,\infty).$$

$$v(t) = S_v(t)v_o + v_1(t) + \int_0^t S_v(t-s)[v(-g+au)](s)\,ds \tag{31b}$$

The Comparison Theorem p.123 implies

$$0 \leq u_1(x,t) \leq \bar{u} \quad \text{and} \quad 0 \leq v_1(x,t) \leq \bar{v} \quad \text{for all } x \in \Omega,\ t \in [0,\infty). \tag{32}$$

Hence the terms u_1, v_1 in the mild formulation (31) are unessential for the derivation of the estimates in Lemma 19 and 20 part I.

We apply Lemma 20 p.72 to the equation (31a) for the component u. Since $u \geq 0$, $v \geq 0$ we can use the one-sided version.

We have weight function $c \equiv 1$ and use (23) as primary a priori estimate. Hence with $q_1 = q_2 = r_2 = \infty$, $r_1 = 1$, $\gamma = 1$ assertion (234) of Lemma 20 p.72 implies (28a) with $p = \infty$. In case $p = 1$, estimate (28a) is just the primary a priori estimate (23). Now (28a) for arbitrary p follows by the interpolation (124) p.51.

Choosing $q_1 = p$ and $q_2 < 1/\delta$ estimate (29) follows easily from (28a).

We apply Lemma 20 p.72 to the equation (31b) for the component v. Consider $c = -g + au$ as a known weight function. For this weight function, (29) and $n/p_0 < 1$ yield the estimate

$$\|c\|_{q_1,q_2} \leq M_1$$

with $q_1,q_2 \in [1,\infty]$ satisfying $n/q_1 + 1/q_2 < 1$. Hence with $r_1 = 1$, $r_2 = \infty$, $\gamma = 1$ assertion (234) of Lemma 20 p.72 implies (28b) with $q = \infty$. In case In case $q = 1$, estimate (28b) is just the primary a priori estimate (23). Now (28b) for arbitrary q follows by interpolation.
Thus the Lemma is proved.

Lemma 3

Let $p_0,q_0 \in [1,\infty)$, $\varepsilon \in (0,1)$ satisfy

$$n/p_0 < 1 - \varepsilon.$$

Take regular convenient initial data.

Then there exists a constant M as specified p.192 such that the solution (u,v) of system (1) satisfies

$$\|u(t) - S_u(t)u_0 - u_1(t)\|_{q_0} \leq M\, t^{\varepsilon} \qquad \text{for all } t\in(0,1); \tag{33a}$$

$$\|v(t) - S_v(t)v_0 - v_1(t)\|_{q_0} \leq M\, t^{\varepsilon} \qquad \text{for all } t\in(0,1). \tag{33b}$$

Here u_1,v_1 are given by (30) in case of Dirichlet boundary conditions (1c) and $u_1 = v_1 = 0$ in case of Neumann boundary conditions (1d).

Proof

With $\delta = n/p_0$ equations (31a)(31b) and Lemma 2 imply the estimate

$$\|u(t) - S_u(t)u_0 - u_1(t)\|_{q_0} + \|v(t) - S_v(t)v_0 - v_1(t)\|_{q_0}$$

$$\leq \int_0^t m(s)^{-\delta} ds\, M_2(1 + \|u\|_{\infty,\delta,t})(1 + \|v\|_{q_0,o,t})$$

$\leq M_3 t^{\varepsilon}$ for all $t\in(0,1)$, which proves the Lemma.

We continue the proof of Theorem 1.

Arbitrary convenient initial data (u_o, v_o) can be approximated by a sequence (u_{om}, v_{om}) of positive regular initial data such that

$$\lim_{m\to\infty} [\|u_{om}-u_o\|_1 + \|v_{om}-v_o\|_1] = 0$$

$$\sup_{m\in N} [\|u_{om}\|_{p_o} + \|\log u_{om}\|_1 + \|v_{om}\|_{q_o} + \|\log v_{om}\|_1] \leq M_3.$$

Hence Lemma 1,2, and 3 above hold with constants M independent of $m \in N$. Now assertions (3a)(3b)(4a)(4b) follow by the arguments of Theorem 2 p.120.

We turn to the investigation of the asymptotic behavior for $t \to \infty$ in the case of Dirichlet boundary conditions (1c). Let Ω_1 be a bounded domain with smooth boundary such that $\overline{\Omega} \subset \mathrm{int}(\Omega_1)$. Let $\varphi \in C^2(\overline{\Omega}_1)$ and $\lambda_1 > 0$ be the principal eigenfunction and eigenvalue of $-\Delta$ in Ω_1:

$$-\Delta\varphi = \lambda_1\varphi \quad \text{in} \quad \Omega_1 \ , \quad \varphi = 0 \quad \text{on} \quad \partial\Omega_1.$$

(See e.g. Amann [4] for details). Since the strong maximum principle implies $\varphi(x) > 0$ for all $x \in \Omega$, we can choose φ such that

$$-\Delta\varphi \geqq 1 \quad \text{and} \quad \varphi \geqq 1 \quad \text{in } \overline{\Omega}.$$

Hence with $\lambda = \lambda_1 \min\{D_u, D_v\}$, formula (20) of Lemma 1 implies

$$d\Lambda_\varphi/dt + \lambda\Lambda_\varphi \leq 0 \qquad \text{for all } t\in[0,\infty)$$

and hence by integration

$$\Lambda(t) \leq \Lambda_\varphi(t) \leq e^{-\lambda t} \|\varphi\|_\infty \Lambda(0) \quad \text{for all } t\in[0,\infty). \tag{34}$$

By Lemma 2 we have the uniform estimate

$$\sup_{1<t<\infty} \|u(t)\|_\infty \leq U. \tag{35}$$

A straightforward calculation yields

$$(u - \overline{u})^2 \leq (2/\overline{u}) L(u,\overline{u}) \quad \text{for all } u\in(0,U]. \tag{36}$$

Now (34)(35)(36) and the definition (19) of $\Lambda(t)$ yield

$$\|u(t) - \overline{u}\|_2 + \|v(t) - \overline{v}\|_2 \leq M_4 e^{-\lambda t/2} \quad \text{for all } t\in[1,\infty). \tag{37}$$

Hence by interpolation we get for all $p\in[2,\infty)$

$$\|u(t)-\overline{u}\|_p \leq \|u(t)-\overline{u}\|_\infty^{1-2/p} \|u(t)-\overline{u}\|_2^{2/p} \leq M_5 e^{-\lambda t/p} \quad \text{for all } t\in[1,\infty) \tag{38}$$

and similarly

$$\|v(t) - \bar{v}\|_p \leq M_6 e^{-\lambda t/p} \quad \text{for all } t\in[1,\infty). \tag{39}$$

Next we derive estimates in Hölder norms. In the following let $0 < t_0 < t_0+t$ and $t_0 < t_0+t_1 < t_0+t_1+t_2$. We use the mild equation

$$u(t_0+t) - \bar{u} = S_u(t)[u(t_0) - \bar{u}] - b\int_0^t S_u(t-s)u(t_0+s)[v(t_0+s) - \bar{v}]\,ds$$
$$\text{for all } t_0, t\in(0,\infty).$$

Let $p\in[2,\infty)$, $\alpha,\mu\in(0,1)$, $\nu\in(0,2)$ satisfy $\alpha + \mu < 1$, $n/p + \nu/2 < \alpha$. Let $A = A_p$ $(= -D_u\Delta)$ denote the generator of the analytic semigroup $S_u(t)$ in the space $L_p(\Omega)$. We estimate:

$$\|A^{\alpha+\mu}[u(t_0+t) - \bar{u}]\|_p \leq \|A^{\alpha+\mu}S_u(t)\|_{p,p}\|u(t_0) - \bar{u}\|_p +$$
$$+ M_7\int_0^t \|A^{\alpha+\mu}S_u(t-s)\|_{p,p}\|u(t_0+s)\|_\infty\|v(t_0+s) - \bar{v}\|_p\,ds$$
$$\text{for all } t_0, t\in(0,\infty).$$

Hence (28a)(38) and the theory of analytic semigroups (Henry [24], Theorem 1.4.3, p.26) imply

$$\|A^{\alpha+\mu}[u(t_0+t) - \bar{u}]\|_p \leq M_8(t^{1-(\alpha+\mu)} + t^{-(\alpha+\mu)})e^{-\lambda t_0/p} \tag{40}$$
$$\text{for all } t_0\in[1,\infty),\ t\in(0,\infty).$$

Concerning Hölder continuity in time we estimate analogously

$$\|A^{\alpha}u(t_0+t_1+t_2) - u(t_0+t_1)\|_p \tag{41}$$
$$\leq \|A^{-\mu}[S_u(t_2) - I]\|_{p,p}\|A^{\alpha+\mu}[u(t_0+t_1) - \bar{u}]\|_p +$$
$$+ \int_0^{t_2} \|A^{\alpha}S_u(t_2-s)\|_{p,p}\|u(t_0+s)\|_\infty\|v(t_0+s) - \bar{v}\|_p\,ds$$
$$\leq M_9(t_2^{\mu} + t_2^{1-\alpha})e^{-\lambda t_0/p} \quad \text{for all } t_0\in[1,\infty),\ t_1,t_2\in(0,\infty).$$

Since the embedding $C^{\nu}(\bar{\Omega}) \subset D(A_p^{\alpha})$ is continuous (Henry [24], Theorem 1.6.1, p.39) estimates (40)(41) show that the function $U(t_0) = U(t_0)(x,t) = u(x,t_0+t)$ satisfies

$$\|u(.,t_0+.) - \bar{u}\|_{C^{\mu}([0,\infty),C^{\nu}(\bar{\Omega}))} \leq M_{10}e^{-\lambda t_0/p} \quad \text{for all } t_0\in[1,\infty). \tag{42}$$

A similar estimate holds for the component v and hence for the right-hand side of system (1).

As explained p.118 the linear Schauder estimates (Ladyzenskaja [32] p.320) imply with $\alpha = \min\{\nu, \mu\}$:

$$\|u(.,t_0+.)-\bar{u}\| + \|v(.,t_0+.)-\bar{v}\|_{C^{2+\alpha,1+\alpha/2}([0,\infty)\times\bar{\Omega})} \leq M_{11}e^{-\lambda t_0/p} \quad \text{for all } t_0\in[1,\infty). \tag{43}$$

Thus (5) follows. Hence Theorem 1 is proved.

Proof of Theorem 2

The set of convenient initial data consists of all $(u_0,v_0) \in L_{p_0}(\Omega) \times L_{q_0}(\Omega)$ as specified in Theorem 2.

The set of regular initial data consists of all $(u_0,v_0) \in C^{2+\alpha}(\bar{\Omega}) \times C^{\alpha}(\bar{\Omega})$ such that $u_0(x) > 0$ and $v_0(x) > 0$ for all $x \in \bar{\Omega}$ and the following compatibility conditions on the boundary:

$$u_0(x) = \bar{u},\ \Delta u_0(x) = 0 \text{ and } v_0(x) = \bar{v} \text{ for all } x \in \partial\Omega. \tag{44}$$

By Theorem 1 p.111 the system (6a)-(6d) has a unique classical solution for regular convenient initial data on some maximal time interval $[0,T_{max})$. The Comparison Theorem p.123 implies (18). Furtheron, it is easy to see that the solution (u,v) grows at most exponentially in time (see Williams and Chow [69]). Hence the explosion property (8) p.112 implies $T_{max} = \infty$.

In the following Lemmas 4,5,6 and 7 we restrict ourselves to regular convenient initial data.

Lemma 4 uses the methods of part I (Lemma 20 p.72) to derive uniform a priori estimates from the primary estimate given by the Lyapunov functional of Lemma 1.

Lemma 5 estimates how the initial data are approximated for $t \to 0$.

Lemma 6 proves continuous dependence of the solution on the initial data.

The asymptotic behavior is investigated p.203-206.

As already stated, we denote by the letter M constants depending only on the exponents, the domain Ω, the constants a,b,f,g and D from the system, the norms $\|u_0\|_{p_0}$, $\|v_0\|_{q_0}$ and the initial value $\Lambda(0)$ of the Lyapunov functional.

Lemma 4

Let $p_o, q_o \in [1,\infty]$ satisfy

$$n/p_o < 1 , \quad n/q_o < 1 .$$

Take regular convenient initial data.

For all $T\in(0,\infty)$ there exists a constant $M(T)$ depending on T and the quantities specified p.192 such that the solution (u,v) of system (6) satisfies the estimates

$$\sup_{0<t<T} m(t)^{\delta}\|u(t)\|_p \le M(T) \quad \text{for all } p\in[p_o,\infty],\ \delta = n/p_o - n/p; \tag{46}$$

$$\sup_{0<t<T} \|v(t)\|_{q_o} \le M(T); \tag{47}$$

$$\sup_{0<t<T} \|L(v(.,t),\overline{v})\|_{q_o} \le \|L(v_o(.),\overline{v})\|_{q_o} + M(T). \tag{48}$$

In the following two cases

(a) space dimension $N = 1$ and lower signs in (6) (diffusing predator)

(b) arbitrary space dimension and upper signs in (6) (diffusing prey)

estimate (46) holds with M independent of T and estimate (47) holds with $q_o = 1$ and M independent of T.

Proof

Take first the case of diffusing prey (upper signs in (6)).

As in the proof of Lemma 2 one shows (46) with $M(T)$ independent of T. To show (47), note that the differential equation (6b) for the component v implies the estimate with $\delta = n/p_o$:

$$\begin{aligned}\|v(t_o+t) - v(t_o)\|_{q_o} &\le a\int_0^t \|(u - \overline{u})(s)\|_{\infty}\|v(s)\|_{q_o}\, ds \\ &\le M_{12}m(t)^{1-\delta}\|u\|_{\infty,\delta,\infty} \sup\{\|v(s)\|_{q_o} \mid s\in[t_o,t_o+t]\} \\ &\le M_{13}m(t)^{1-\delta} \sup \|v(s)\|_{q_o} \quad \text{for all } t_o,t\in[0,\infty).\end{aligned} \tag{49}$$

Now choose $h\in(0,1)$ small enough such that $M_{12}m(h)^{1-\delta} \le 1/2$.
Then (49) implies

$$\sup\{\|v(s)\|_{q_o} \mid s \in [t_o,t_o+h]\} \le 2\|v(t_o)\|_{q_o} \quad \text{for all } t_o\in[0,\infty).$$

Now (47) follows by induction.

Take the case of diffusing predator (lower signs).

At first, we consider the component v. Since $u \geqq 0$, $v \geqq 0$, the differential equation (6b) implies

$$v_t \leq gv \qquad \text{for all } x \in \Omega,\ t\in[0,\infty)$$

and hence by integration

$$v(x,t) \leq e^{gt}v_o(x) \quad \text{for all } x \in \Omega,\ t\in[0,\infty).$$

Now consider the component u. In the differential equation (6a) for u, we consider $c = bv - f$ on the right-hand side as a known weight function. We apply Lemma 20 p.72. We restrict all considerations to an arbitrary interval $[0,T]$ and take (23) as a primary a priori estimate for u and (50) as an estimate for the weight function. Hence in Lemma 20 there appear $q_1 := q_o$, $r_1 := 1$, $q_2 := r_2 := \infty$, $\gamma := 1$. Assertion (234) of Lemma 20 shows (46) with $p = \infty$.
In case (a) of space dimension $N = 1$, we get a better result. We use (23) as a primary a priori estimate for u as well as an estimate for the weight function. Hence in Lemma 20 there appear $q_1 = r_1 = 1$, $q_2 = r_2 = \infty$, $\gamma = 1$. Thus assumption (231) can only be fulfilled for space dimension $N = 1$, $n = 1/2$. Assertion (234) shows (46) with $p = \infty$ and $M(T)$ independent of T.
For arbitrary p, estimate (46) follows by interpolation (124) p.51.

It remains to show (48). The differential equation (6b) for v implies

$$\frac{d}{dt}L(v(x,t),\bar{v}) = \pm a(v(x,t) - \bar{v})(u(x,t) - \bar{u}) \quad \text{for all } x \in \Omega,\ t\in[0,\infty).$$

After integrating we get the estimate with $\delta = n/p_o$:

$$\|L(v(.,t),\bar{v})\|_{q_o} \leq$$

$$\leq \|L(v_o(.),\bar{v})\|_{q_o} + M_{14}(t^{1-\delta} + t)(1+\|u\|_{\infty,\delta,t})(1+\|v\|_{q_o,o,t})$$

for all $t\in[0,\infty)$.

Now (46)(47) imply the assertion (48). Thus the Lemma is proved.

<u>Lemma 5</u>

Let $p_o, q_o \in [1,\infty]$, $\varepsilon \in (0,1)$ satisfy

$$n/p_o < 1\ ,\quad n/q_o < 1\ ,\quad \varepsilon = \min\{1-n/p_o, 1-n/q_o\}\ . \tag{50}$$

Take regular convenient initial data.

Then there exists a constant M as specified p.192 such that the solution (u,v) of system (6) satisfies

$$\|u(t) - S(t)u_o - u_1(t)\|_{p_o} \leq M\, t^{\varepsilon} \quad \text{for all } t\in(0,1); \tag{51a}$$

$$\|v(t) - v_o\|_{q_o} \leq M\, t^{\varepsilon} \quad \text{for all } t\in(0,1). \tag{51b}$$

Here $u_1(t)$ is defined by (30a).

<u>Proof</u>

With $\delta = n/p_o$ equation (31a) leads to the estimate

$$\|u(t) - S(t)u_o - u_1(t)\|_{p_o} \leq$$

$$\leq M_{15}\int_0^t \|S(t-s)\|_{p_o,q_o} m(s)^{-\delta} ds\ (1+\|u\|_{\infty,\delta,t})(1+\|v\|_{q_o,o,t})$$

for all $t\in(0,\infty)$.

To estimate the integral we apply Lemma 6 p.31 with $f \equiv 1$ and

$p:= p_o,\ s_1:= q_o,\ s_2:= \infty,\ \alpha:= n/p_o,\ \varepsilon:= \varepsilon,\ \delta:= 0.$

To estimate the norms of u and v, apply (46)(47) of Lemma 4 above. Hence we get (51a). As in the proof of Lemma 3 p.195 one gets (51b). Thus the Lemma is proved.

<u>Lemma 6</u>

Let $p_o,q_o\in[1,\infty]$, $\delta\in[0,1)$ satisfy

$$n/p_o = \delta < 1\ ,\quad n/q_o < 1\ .$$

For i = 1,2 let (u_i,v_i) be the solutions of system (6) for the regular convenient initial data (u_{oi},v_{oi}).

For all $T\in(0,\infty)$ there exists a constant M(T) depending on T and the quantities specified p.192 such that

$$\|u_1-u_2\|_{\infty,\delta,T} + \|v_1-v_2\|_{q_o,o,T} \leq M(T)[\|u_{o1}-u_{o2}\|_{p_o} + \|v_{o1}-v_{o2}\|_{q_o}]. \tag{52}$$

<u>Proof</u>

Let $\varepsilon = \min\{1-n/p_o, 1-n/q_o\}$ and $T\in(0,\infty)$ be arbitrary.

The solution (u,v) of system (6) satisfies the integral equations (31a)(31b) with $S_u(t) = S(t)$ and $S_v(t) = I$ (identity).

Subtracting the integral equations (31a) corresponding to i = 1,2 yields the estimate

$$\|(u_1-u_2)(t)\|_\infty \le \|S(t)\|_{\infty,p_o}\|u_{o1}-u_{o2}\|_{p_o} +$$

$$+ M_{16}\int_0^t \|S(t-s)\|_{\infty,q_o} m(s)^{-\delta}ds\, B(u_i,v_i) \quad \text{with}$$

$$B(u_i,v_i) = \|u_1-u_2\|_{\infty,\delta,t}(1+\|v_1\|_{q_o,o,t}) + (1+\|u_2\|_{\infty,\delta,t})\|v_1-v_2\|_{q_o,o,t}.$$

Hence Lemma 6 p.31 can be used to estimate the integral and (46)(47) can be used to estimate the norms of v_1 and u_2. Thus we get

$$\|u_1-u_2\|_{\infty,\delta,t} \le K\|u_{o1}-u_{o2}\|_{p_o} + M_{17}(T)t^{\varepsilon}[\|u_1-u_2\|_{\infty,\delta,t} + \|v_1-v_2\|_{q_o,o,t}]$$

$$\text{for all } t\in(0,T]. \qquad (54)$$

Subtracting the integral equations (31b) for i = 1,2 one gets similarly

$$\|(v_1-v_2)(t)\|_{q_o} \le \|v_{o1}-v_{o2}\|_{q_o} + M_{18}\int_0^t m(s)^{-\delta}ds\, B(u_i,v_i)$$

and hence

$$\|v_1-v_2\|_{q_o,o,t} \le \|v_{o1}-v_{o2}\|_{q_o} + M_{19}(T)t^{\varepsilon}[\|u_1-u_2\|_{\infty,\delta,t} + \|v_1-v_2\|_{q_o,o,t}]$$

$$\text{for all } t\in(0,T]. \qquad (55)$$

Choose $h\in(0,1)$ small enough such that $M_{17}h^{\varepsilon} \le 1/2$, $M_{19}h^{\varepsilon} \le 1/2$. Then (54)(55) imply

$$\|u_1-u_2\|_{\infty,\delta,h} + \|v_1-v_2\|_{q_o,o,h} \le (2K+1)[\|u_{o1}-u_{o2}\|_{p_o} + \|v_{o1}-v_{o2}\|_{q_o}].$$

The same proof shows that even

$$\|(u_1-u_2)(.,t_o+.)\|_{\infty,\delta,h} + \|(v_1-v_2)(.,t_o+.)\|_{q_o,o,h} \le$$

$$\le (2K+1)[\|(u_1-u_2)(t_o)\|_{p_o} + \|(v_1-v_2)(t_o)\|_{q_o}] \quad \text{for all } t_o\in[0,T).$$

Hence the assertion (52) follows by iteration. Thus the Lemma is proved.

We continue the proof of Theorem 2.
Arbitrary convenient initial data (u_o,v_o) as assumed in Theorem 2 can be approximated by a sequence (u_{om},v_{om}) of positive regular initial data such that

$$\lim_{m\to\infty} [\|u_{om}-u_o\|_{p_o} + \|v_{om}-v_o\|_{q_o}] = 0, \qquad (56)$$

$$\sup_{m\in N} [\|u_{om}\|_{p_o} + \|\log u_{om}\|_1 + \|v_{om}\|_{q_o} + \|\log v_{om}\|_1] \le M_3. \qquad (57)$$

Let (u_m, v_m) be the global solution of system (6) for initial data (u_{om}, v_{om}). The constants M in Lemma 1,4,5 and 6 can be chosen independent of $m \in N$. Hence assertion (52) of Lemma 6 implies that the sequence (u_m, v_m) converges to $(u,v) \in E_{\infty,\delta,T} \times E_{q_o,o,T}$ for all $T\in(0,\infty)$ (taking convenient restrictions of u and v).
Estimates (51a)(51b) hold for the limit (u,v). Hence we get (8a)(8b). (We exclude $p_o = \infty$ or $q_o = \infty$ in order to guarantee (56) without caring the compatibility assumptions like (44)).
By the arguments p.122 it is straightforward that (u,v) is indeed a mild solution of system (6) for initial data (u_o, v_o).

We turn to the investigation of the asymptotic behavior.

Assertions (9)(10)(11) follow by Lemma 1. Furtheron, we consider the cases (a)(b) as specified in the Theorem p.190. Assertion (12) is proved in Lemma 4. It remains to show (13)-(16).

By (6a) the function $L = L(u(x,t),\bar{u})$ satisfies the differential equation

$$\left[\frac{\partial}{\partial t} - D\Delta\right] L(u,\bar{u}) = \mp\, b(u-\bar{u})(v-\bar{v}) - D(\nabla \log u)^2 \quad \text{for all } x \in \Omega,\ t\in(0,\infty).$$

Let $\varphi \in C^2(\bar{\Omega})$ be an arbitrary test function. Since $L = \partial L/\partial n = 0$ on the boundary $\partial\Omega$, we get by spatial integration

$$\frac{d}{dt}\int_\Omega \varphi L(u,\bar{u})dx - D\int_\Omega (\Delta\varphi) L(u,\bar{u})dx =$$

$$\mp\int_\Omega \varphi(u-\bar{u})(v-\bar{v})dx - D\int_\Omega \varphi(\nabla\log u)^2 dx \quad \text{for all } t\in(0,\infty).$$

Finally we get by integration over the time interval $[t,t+T] \subset (0,\infty)$

$$\int_\Omega \varphi(x)L(u(x,t+T),\bar{u})dx - \int_\Omega \varphi(x)L(u(x,t),\bar{u})dx = \tag{58}$$

$$= D\int_t^{t+T}\int_\Omega (\Delta\varphi)L(u,\bar{u}) - \varphi(\nabla\log u)^2\, dxd\tau\ \mp$$

$$\mp\int_t^{t+T}\int_\Omega \varphi(u-\bar{u})(v-\bar{v})\, dxd\tau.$$

The right-hand side can be estimated by (22a)(23)(46).
Hence we get for the functional

$\Sigma: t\in(0,\infty) \to \Sigma(t)\in[0,\infty)$ given by

$$\Sigma(t) = \int_\Omega L(u(x,t),\bar{u})dx$$

that

$$|\Sigma(t+T) - \Sigma(t)| \leq M_{20}T \quad \text{for all } t\in[1,\infty),\ T\in[0,\infty). \tag{59}$$

By the construction p.196, there exists a test function $\varphi \in C^2(\overline{\Omega})$ such that $\varphi \geqq 1$ and $-\Delta\varphi \geqq 1$ in the domain Ω. Hence (22a) of Lemma 1 implies

$$\int_1^\infty \Sigma(t)dt \leqq M_{21} < \infty. \tag{60}$$

Now (59)(60) imply

$$\lim_{t\to\infty} \Sigma(t) = 0 \tag{61}$$

proving (14a) and by Jensens's inequality (27) assertion (13a), too.

We turn to the component v. We take the limit $T \to \infty$ in (58) and use (22a) of Lemma 1 and (61). Hence we get

$$\int_1^\infty\int_\Omega \varphi(x)(u(x,t)-\overline{u})(v(x,t)-\overline{v})\;dxdt \leqq M_{22}\|\varphi\|_{C^2} \quad \text{for all } \varphi \in C^2(\overline{\Omega}). \tag{62}$$

By (6a) the function $w = \log(u(x,t)/\overline{u})$ satisfies the differential equation

$$\left[\frac{\partial}{\partial t} - D\Delta\right]\log u/\overline{u} = \mp b(v-\overline{v}) + D(\nabla\log u)^2 \quad \text{for all } x \in \Omega, t\in(0,\infty).$$

Take a test function $\varphi \in C_0^2(\overline{\Omega})$. Integration over the domain Ω and an arbitrary time interval $[t,t+T] \subset (0,\infty)$ yields

$$\begin{aligned}\int_\Omega \varphi(x)\log(u(x,t+T)/\overline{u})dx - \int_\Omega \varphi(x)\log(u(x,t)/\overline{u})dx = \\ = D\int_t^{t+T}\int_\Omega (\Delta\varphi)\log u/\overline{u} - \varphi(\nabla\log u)^2\;dxd\tau \mp \\ \mp b\int_t^{t+T}\int_\Omega \varphi(v-\overline{v})\;dxd\tau.\end{aligned} \tag{63}$$

Define the functional $V_\varphi\colon t\in(0,\infty) \to V_\varphi(t)\in R$ by

$$V_\varphi(t) = \int_\Omega \varphi(x)(v(x,t)-\overline{v})dx.$$

In (63) we fix T and take the limit $t \to \infty$. By (12)(13a)(14a) and (22a) the first and second line in (63) converge to zero. Thus we get

$$\lim_{t\to\infty} \int_t^{t+T} V_\varphi(\tau)d\tau = 0 \quad \text{for all } T\in(0,\infty), \text{ all } \varphi \in C_0^2(\overline{\Omega}). \tag{64}$$

The differential equation (6b) for the component v implies

$$\frac{d}{dt} V_\varphi(t) = \mp a\int_\Omega \varphi(u-\overline{u})(v-\overline{v})dx \pm a\overline{v}\int_\Omega \varphi(u-\overline{u})dx \quad \text{for all } t\in(0,\infty).$$

By (13a) the second term on the right-hand side of (65) tends to zero for $t \to \infty$. After integration with respect to time, the first term on the right-hand side of (65) can be handled by (62). Hence we get

$$\lim_{t\to\infty} |V_\varphi(t) - V_\varphi(t+T)| = 0 \quad \text{for all } T\in(0,\infty), \text{ all } \varphi \in C_0^2(\overline{\Omega}). \tag{66}$$

Now (64) and (66) imply

$$\lim_{t\to\infty} V_\varphi(t) = 0 \qquad \text{for all } \varphi \in C_0^2(\overline{\Omega}). \tag{67}$$

Since by (9)

$$\sup_{1<t<\infty} \|v(t)\|_1 \leq M_{23},$$

a standard density argument shows that

$$\lim_{t\to\infty} V_\varphi(t) = 0 \qquad \text{for all } \varphi \in C_0(\overline{\Omega}),$$

which proves (13b).

To show (14b), consider at first a test function $\varphi \in C^{2+\alpha}(\overline{\Omega})$ such that $\varphi \geq 0$ and $-\Delta\varphi \geq 0$. By (21) of Lemma 1 the Lyapunov functional $\Lambda_\varphi(t)$ converges to some limit for $t \to \infty$. By (61) the part coming from the component u converges to zero. Hence the limit in (14b),

$$\lim_{t\to\infty} \int_\Omega \varphi(x) L(v(x,t),\overline{v})dx, \text{ exists for all } \varphi \in C^{2+\alpha}(\overline{\Omega}) \text{ with } \varphi \geq 0, -\Delta\varphi \geq 0.$$

Arbitrary $\varphi \in C^{2+\alpha}(\overline{\Omega})$ can be decomposed as $\varphi = \varphi_1 - \varphi_2$ such that $\varphi_i \in C^{2+\alpha}$ and $\varphi_i \geq 0$, $-\Delta\varphi_i \geq 0$ for $i = 1,2$.
Hence the limit in (14b) exists for all $\varphi \in C^{2+\alpha}(\overline{\Omega})$. Since

$$\sup_{1<t<\infty} \int_\Omega L(v(x,t),v)dx \leq M_{24},$$

a standard density argument shows that the limit in (14b) exists for all $\varphi \in C(\overline{\Omega})$.

It remains to show (15)(16) in case of space dimension $N = 1$. By (46)(47) of Lemma 4 we have

$$\sup_{1<t<\infty} [\|u(t)\|_\infty + \|v(t)\|_1] \leq M_{25}.$$

Applying Lemma 21 p.78 to the equation (6a) for the component u, we get

$$\sup_{1<t<\infty} \|u(.,t+.)\|_{C^\mu([0,\infty),C^\nu(\overline{\Omega}))} \leq M_{26} \quad \text{for all } \mu,\nu\in(0,1) \text{ with } \mu + \nu/2 < 1/2. \tag{68}$$

For $\mu < \hat{\mu}$, $\nu < \hat{\nu}$ the embedding $C^{\hat{\mu}}([0,\infty), C^{\hat{\nu}}(\bar{\Omega})) \subset C^{\mu}([0,\infty), C^{\nu}(\bar{\Omega}))$ is compact. Hence the L_1-convergence of the component u given by (13a) and (68) imply that the component u converges in the Hölder spaces occuring in (68), too. Thus (15) is proved.

It remains to prove (16). Let $\alpha, \mu \in (0,1)$ satisfy $\alpha + \mu < 3/4$. Estimates (272)(274) p.80 in the proof of Lemma 21 (with $n = 1/2$, $q_1 = 1$, $q_2 = \infty$, $p = 2$) show that

$$\sup_{1<t<\infty} \|u(t)\|_{C^{\mu}([0,\infty), D(A_2^{\alpha}))} \leq M_{27}.$$

Since for $1/2 < \alpha$, the embedding $D(A_2^{\alpha}) \subset H_1(\Omega)$ is compact, convergence of the component u in the weaker sense above implies convergence in the Hilbert space $H_1(\Omega)$. Thus (16) is proved.

The proof of Theorem 2 is finished.

In the rest of this section, we exploit a Lyapunov functional to investigate a slightly generalized Volterra-Lotka system.

Let the functions $a, b, f, g, h, k \in C^1([0,\infty))$ satisfy the following assumptions:

$$a'(u) \geq 0 \quad \text{and} \quad b'(u) \geq 0 \quad \text{for all } u \in [0,\infty) \tag{ab}$$

$$f'(u) \leq 0 \quad \text{and} \quad g'(u) \geq 0 \quad \text{for all } u \in [0,\infty) \tag{fg}$$

$$h(u) > 0 \quad \text{and} \quad k(u) > 0 \quad \text{for all } u \in (0,\infty) \; ; \quad h(0) = k(0) = 0 \tag{hk}$$

There exists a constant $M \in (0,\infty)$ such that (h) and (k1) or (k2) hold

$$h(u)[f(u) - a(0)] \leq M(1+u) \quad \text{for all } u \in [0,\infty) \tag{h}$$

$$k(v) \leq M(1+v) \quad \text{for all } v \in [0,\infty) \tag{k1}$$

$$k(v)[-g(v) + b_{\infty}] \leq M(1+v) \quad \text{for all } v \in [0,\infty) \text{ with } b_{\infty} = \lim_{u\to\infty} b(u) \tag{k2}$$

There exists a unique equilibrium $(\bar{u}, \bar{v}) \in (0,\infty) \times (0,\infty)$ satisfying

$$f(\bar{u}) = a(\bar{v}) \quad \text{and} \quad g(\bar{v}) = b(\bar{u}) \; ; \quad a'(\bar{v}) > 0 \; ; \quad b'(\bar{u}) > 0 \tag{eq}$$

$$\frac{d}{du} \frac{b(u) - b(\bar{u})}{h(u)} \geq 0 \quad \text{for all } u \in (0,\infty) \tag{bh}$$

$$\frac{d}{dv} \frac{a(v) - a(\bar{v})}{k(v)} \geq 0 \quad \text{for all } v \in (0,\infty) \tag{ak}$$

We consider the generalized Volterra-Lotka system

$$u_t - D_u \Delta u = h(u)[f(u) - a(v)] \qquad \text{for all } x \in \Omega\ ,\ t > 0 \tag{69a}$$

$$v_t - D_v \Delta v = k(v)[-g(v) + b(u)] \tag{69b}$$

with Dirichlet boundary conditions

$$u(x,t) = \bar{u} \quad \text{and} \quad v(x,t) = \bar{v} \qquad \text{for all } x \in \partial\Omega,\ t > 0 \tag{69c}$$

and initial conditions

$$u(x,0) = u_o(x) \quad \text{and} \quad v(x,0) = v_o(x) \quad \text{for all } x \in \Omega. \tag{69d}$$

<u>Theorem 3</u> (Boundedness and convergence to equilibrium for some generalized Volterra-Lotka systems)

Take initial data $(u_o, v_o) \in C^{2+\alpha}(\bar{\Omega}) \times C^{2+\alpha}(\bar{\Omega})$ satisfying

$$u_o(x) > 0 \quad \text{and} \quad v_o(x) > 0 \qquad \text{for all } x \in \bar{\Omega};$$

$$u_o(x) = \bar{u}\ ,\ \Delta u_o(x) = 0 \quad \text{and} \quad v_o(x) = \bar{v}\ ,\ \Delta v_o(x) = 0 \quad \text{for all } x \in \partial\Omega.$$

(These are the compatibility conditions of first order).

Then the generalized Volterra-Lotka system (69a)-(69d) has a global classical solution (u,v) on the time interval $[0,\infty)$.

There exist $\lambda, M \in (0,\infty)$ such that this solution satisfies

$$\|u(.,t) - \bar{u}\|_{C^2(\bar{\Omega})} + \|v(.,t) - \bar{v}\|_{C^2(\bar{\Omega})} \leq M\, e^{-\lambda t} \quad \text{for all } t \in (0,\infty). \tag{70}$$

<u>Proof</u>

The existence of a classical solution (u,v) on some maximal interval $[0,T_{max})$ follows by Theorem 1 p.111. The Comparison Theorem p.123 shows

$$u(x,t) > 0 \quad \text{and} \quad v(x,t) > 0 \qquad \text{for all } x \in \bar{\Omega},\ t \in [0,T_{max}). \tag{71}$$

We define the functions $A, B \in C^2(0,\infty)$ by

$$B(\bar{u}) = 0\ ,\quad B'(u) = \frac{b(u) - b(\bar{u})}{h(u)} \qquad \text{for all } u \in (0,\infty);$$

$$A(\bar{v}) = 0\ ,\quad A'(v) = \frac{a(v) - a(\bar{v})}{k(v)} \qquad \text{for all } v \in (0,\infty).$$

By the assumptions above the functions A,B have the following properties:

$$\begin{aligned}
&B(u) \geq 0\ ,\quad B''(u) \geq 0 \quad \text{for all } u\in(0,\infty);\\
&A(v) \geq 0\ ,\quad A''(v) \geq 0 \quad \text{for all } v\in(0,\infty);\\
&B(\bar{u}) = B'(\bar{u}) = 0\ ,\quad B''(\bar{u}) > 0\ ,\quad \lim_{u\to\infty} B(u) = \infty;\\
&A(\bar{v}) = A'(\bar{v}) = 0\ ,\quad A''(\bar{v}) > 0\ ,\quad \lim_{v\to\infty} A(v) = \infty.
\end{aligned} \tag{72}$$

Let $\varphi \in C^2(\bar{\Omega})$ be an arbitrary test function and (u,v) the solution of system (69). Define the functional

$\Lambda_\varphi: t\in[0,\infty) \to \Lambda_\varphi(t)\in R$ by setting

$$\Lambda_\varphi(t) = \int_\Omega \varphi(x)\,[B(u(x,t)) + A(v(x,t))]\ dx. \tag{73}$$

Lemma 7

$$\begin{aligned}
d\Lambda_\varphi/dt = &\int_\Omega (\Delta\varphi)\,[D_u B(u) + D_v A(v)]\ dx - \\
&- \int_\Omega \varphi[D_u B''(u)(\nabla u)^2 + D_v A''(v)(\nabla v)^2]\ dx + \\
&+ \int_\Omega \varphi[b(u)-b(\bar{u})][f(u)-f(\bar{u})] - \varphi[a(v)-a(\bar{v})][g(v)-g(\bar{v})]\ dx
\end{aligned} \tag{74}$$

Take test function $\varphi \equiv 1$. There exist $\lambda, M\in(0,\infty)$ such that

$$0 \leq \Lambda_1(t) \leq M\, e^{-\lambda t} \quad \text{for all } t\in[0,T_{max}). \tag{75}$$

Proof

Differentiating Λ_φ and using the differential equations (69a)(69b) yields

$$\begin{aligned}
d\Lambda_\varphi/dt = &\int_\Omega \varphi B'(u)h(u)[f(u)-a(v)] + \varphi A'(v)k(v)[-g(v)+b(u)]\ dx + \\
&+ \int_\Omega \varphi D_u B'(u)(\Delta u) + \varphi D_v A'(v)(\Delta v)\ dx = \\
= &\int_\Omega \varphi[b(u)-b(\bar{u})][f(u)-a(v)] + \varphi[a(v)-a(\bar{v})][-g(v)+b(u)]\ dx - \\
&- \int_\Omega \varphi[b(u)-b(\bar{u})][f(\bar{u})-a(\bar{v})] + \varphi[a(v)-a(\bar{v})][-g(\bar{v})+b(\bar{u})]\ dx - \\
&- \int_\Omega D_u\nabla(\varphi B'(u))\ \nabla u + D_v\nabla(\varphi A'(v))\ \nabla v\ dx + \\
&+ \int_{\partial\Omega} D_u\varphi B'(u)\ \frac{\partial u}{\partial n} + D_v\varphi A'(v)\ \frac{\partial v}{\partial n}\ dx\ .
\end{aligned}$$

The boundary integral in the last line vanishes, since by (69c) and (72) we have $B'(u) = B'(\bar{u}) = 0$ and $A'(v) = A'(\bar{v}) = 0$ on $\partial\Omega$.
We get furtheron

$$d\Lambda_\varphi/dt = \int_\Omega \varphi[b(u)-b(\bar{u})][f(u)-f(\bar{u})] - \varphi[a(v)-a(\bar{v})][g(v)-g(\bar{v})]\, dx -$$

$$- \int_\Omega D_u\varphi\, B''(u)\,(\nabla u)^2 + D_v\varphi\, A''(v)\,(\nabla v)^2\, dx -$$

$$- \int_\Omega D_u[\nabla\varphi][\nabla B(u)] + D_v[\nabla\varphi][\nabla A(v)]\, dx \, .$$

One further partial integration of the last line yields (74). Since $B(u) = B(\bar{u}) = 0$ and $A(v) = A(\bar{v}) = 0$ on $\partial\Omega$, no boundary terms occur.

By the construction p.196 there exists $\varphi \in C^2(\bar{\Omega})$ and $\lambda_1 > 0$ such that

$$\varphi \geqq 1 \, , \quad -\Delta\varphi \geqq 1 \quad \text{and} \quad -\Delta\varphi = \lambda_1\varphi \quad \text{in} \quad \bar{\Omega}.$$

Let $\lambda = \lambda_1 \min\{D_u, D_v\}$. By the monotonicity assumptions (ab)(fg) the last line in formula (74) is nonpositive. Hence we get from (72) and (74)

$$d\Lambda_\varphi/dt + \lambda\Lambda_\varphi \leqq 0 \qquad \text{for all } t\in[0,T_{max}).$$

Integration with respect to time yields

$$\Lambda_1(t) \leqq \Lambda_\varphi(t) \leqq \Lambda_\varphi(0)\, e^{-\lambda t} \quad \text{for all } t\in[0,T_{max}),$$

which proves (75). Thus the Lemma is proved.

Lemma 8

There exist $U,V\in(0,\infty)$ such that the solution (u,v) of system (69) satisfies

$$\|u(.,t)\|_\infty \leqq U \quad \text{and} \quad \|v(.,t)\|_\infty \leqq V \quad \text{for all } t\in[0,T_{max}). \qquad (76)(77)$$

Proof

For the component u, the differential equation (69a) and estimates (71) and (h) imply

$$u_t - D_u\Delta u \leqq M(1+u) \quad \text{for all } x \in \Omega,\ t\in[0,T_{max}).$$

We apply Lemma 26 p.100 in the one-sided version with (75) as primary a priori estimate. Since $\lim_{u\to\infty} B(u) = \infty$, the Lyapunov functional $\Lambda_1(t)$ provides indeed an estimate (371) assumed in Lemma 26. Hence assertion (373) of Lemma 26 yields (76).

Now consider the component v. The differential equation (69b) and estimates (71) and (k1) together with (76) or (k2) imply

$$v_t - D_v \Delta v \leq M(1+v)\max\{1,-g(0)+b(U)\} \quad \text{for all } x \in \Omega,\ t \in [0,T_{max}).$$

We apply Lemma 26 p.100 in the one-sided version with (75) as primary a priori estimate. Indeed, since $\lim_{v\to\infty} A(v) = \infty$, estimate (75) can be used as primary estimate. Hence assertion (373) of Lemma 26 yields (77). Thus the Lemma is proved.

We finish the proof of the Theorem. Estimates (76)(77) and the explosion property (8) p.112 show that $T_{max} = \infty$. Thus (u,v) is indeed a global solution on the interval $[0,\infty)$.

We consider the asymptotic behavior. By assumption (72) there exists a constant $C \in (0,\infty)$ such that

$$(u-\bar{u})^2 + (v-\bar{v})^2 \leq C[A(v) + B(u)] \quad \text{for all } u \in (0,U],\ v \in (0,V].$$

Hence (75)(76)(77) imply that there exists $C_1 \in (0,\infty)$ such that

$$\|u(.,t) - \bar{u}\|_2 + \|v(.,t) - \bar{v}\|_2 \leq C_1 e^{-\lambda t/2} \quad \text{for all } t \in [0,\infty). \tag{78}$$

From this estimate corresponding to (37) p.196 we get estimates in Hölder norms and finally the classical Schauder estimate by the same argument as in the proof of Theorem 1 p.197.

Thus Theorem 3 is proved.

References

[1] N.D. Alikakos, An application of the invariance principle to reaction-diffusion equations, J. Diff. Eq., 33 (1979) 201-226.

[2] N.D. Alikakos, L_p-bounds of solutions of reaction-diffusion equations, Comm. Part. Diff. Eq., 4(8) (1979) 827-868.

[3] H. Amann, Invariant sets and existence theorems for semilinear parabolic and elliptic systems, J. Math. Anal. Appl., 65 (1978) 432-467.

[4] H. Amann, Dual semigroups and second order linear elliptic boundary value problems, preprint 1983, to appear in Israel J. Math.

[5] J.F.G. Auchmuty and G. Nicolis, Bifurcation analysis of nonlinear reaction-diffusion equations I: Evolution equations and the steady state solutions, Bull. Math. Biol., 37 (1975) 323-365.

[6] J.M. Ball, Remarks on blow-up and nonexistence theorems for non-linear evolution equations, Quart. J. Math. Oxford, (2), 28 (1977) 473-486.

[7] G.A. Carpenter, A geometric approach to singular perturbation problems with applications to nerve impulse equations, J. Diff. Eq., 23 (1977) 335-367.

[8] K. Chueh, C. Conley and S. Smoller, Positively invariant regions for systems of nonlinear parabolic equations, Indiana Univ. Math. J., 26 (1977) 373-392.

[9] E.D. Conway and J.A.Smoller, Diffusion and the classical ecological interactions, in Nonlinear Diffusion, Research Notes in Math., 14 (1977) 53-69.

[10] O. Diekmann and N.M. Temme, Nonlinear Diffusion Problems, Mathematisch Centrum Amsterdam 1976.

[11] W. Ebel, Lösungen bei Reaktions-Diffusions-Modellen für den träger-vermittelten erleichterten Transport durch biologische Membranen, Dissertation, Universität Tübingen, 1983.

[12] J.W. Evans and J. Feroe, Local stability theory of the nerve impulse, Math. Bios., 37 (1977) 23-50.

[13] P.C. Fife, Mathematical Aspects of Reacting and Diffusing Systems, Lecture Notes in Biomathematics 28, Springer-Verlag, New York 1979.

[14] D.D. Figuereido, P.L. Lions and R.D. Nussbaum, Estimations a priori pour les solutions positive de problèmes elliptiques superlinéares, C. R. Acad. Sc. Paris, 290A (1980) 217-221.

[15] R. Fiorenza, Sui problemi di derivata oblique per le equasioni ellittiche, Ric. Mat., 8(1) (1959) 83-110.

[16] R.A. Fisher, The advance of advantageous genes, Ann. Eugenics, 7 (1937) 355-369.

[17] W.E. Fitzgibbon and H.F. Walker, Nonlinear Diffusion, Research Notes in Math. 14, Pitman, London 1977.

[18] A. Friedman, Partial Differential Equations, Holt, Rinehart and Winston, Inc. 1969.

[19] A. Friedman, Partial Differential Equations of Parabolic Type, Prentice-Hall, Englewood Cliffs, New York 1964.

[20] A. Friedman, Remarks on nonlinear parabolic equations, Proc. Symp. Appl. Math., 17 (1965) 3-23.

[21] K.P. Hadeler, F. Rothe and H. Vogt, Stationary solutions of reaction-diffusion equations, Math. Meth. Appl. Sci., 1 (1979) 418-431.

[22] A. Haraux and F.B. Weissler, Nonuniqueness for a semilinear initial value problem, preprint.

[23] U. an der Heiden, Analysis of Neural Networks, Lecture Notes in Biomathematics 35, Springer-Verlag, New York 1980.

[24] D. Henry, Geometric Theory of Semilinear Parabolic Equations, Lecture Notes in Math. 840, Springer-Verlag, New York 1981.

[25] M. Herschkowitz-Kaufmann, Bifurcation analysis of nonlinear reaction-diffusion equations II: Steady state solutions and comparison with numerical simulations, Bull. Math. Biol., 37 (1975) 589-635.

[26] W.E. Kastenberg and P.L. Chambré, On the stability of nonlinear space-dependent reactor kinetics, Nucl. Sci. Eng., 31 (1968) 67-79.

[27] K. Kawasaki and E. Teramoto, Spatial pattern formation for prey-predator populations, J. Math. Biol., 8 (1979) 33-46.

[28] E.F. Keller and G.M. Odell, Travelling bands of chemotactic bacteria revisited, J. Theor. Biol., 56 (1976) 243-247.

[29] H. Kielhöfer, Existenz und Regularität von Lösungen semilinearer parabolischer Anfangs-Randwertprobleme, Math. Z., 142 (1975) 131-160.

[30] H. Kielhöfer, Global solutions of semilinear evolution equations satisfying an energy inequality, J. Diff. Eq., 36 (1980) 188-222.

[31] H.J. Kuiper, Existence and comparison theorems for nonlinear diffusion systems, J. Math. Anal. Appl., 60 (1977) 166-181.

[32] O.A. Ladyzenskaja, V.A. Solonnikov and N.N. Ural'ceva, Linear and Quasilinear Equations of Parabolic Type, Am. Math. Soc., Providence, Rhode Island (1968), Transl. of Math. Monographs 23.

[33] H. Lange, Die globale Lösbarkeit einer nichtlinearen Reaktionsgleichung, Math. Nachrichten, 80 (1971) 165-181.

[34] O. Lopes, FitzHugh-Nagumo system: boundedness and convergence to equilibrium, Univ. Estadual de Campinas, preprint, to appear in J. Diff. Eq.

[35] K. Maginu, Reaction-diffusion equations describing morphogenesis I: Wave form stability of stationary wave solutions in a one-dimensional model, Math. Biosciences, 27 (1975) 17-98.

[36] P. Massatt, Obtaining L_∞-bounds from L_p-bounds for solutions of various parabolic partial differential equations, Conference Proc., Intern. Symposium on Dynamical Systems, Univ. of Florida 1981.

[37] R. May, Stability and Complexity in Model Ecosystems, Princeton Univ. Press, Princeton, New Jersey 1974.

[38] H. Meinhardt, Models of Biological Pattern Formation, Academic Press, London New York 1982.

[39] H. Meinhardt and A. Gierer, Generation and regeneration of sequences of structures during morphogenesis, J. Theor. Biol., 85 (1980) 429-450.

[40] M. Mimura and J.D. Murray, On a diffusive prey-predator model which exhibits patchiness, J. Theor. Biol., 75 (1979) 249-262.

[41] M. Mimura and Y.Nishiura, Spatial pattern for interaction-diffusion equations in biology, Proc. Intern. Symposium on Math. Topics in Biology, 1978, 136-146.

[42] P. de Mottoni and A. Tesei, Asymptotic stability results for a system of quasilinear parabolic equations, Applicable Analysis, 9 (1979) 7-21.

[43] P. de Mottoni, Qualitative analysis for some quasilinear parabolic systems, Institue of Math., Polish Academy of Sciences, preprint 171, January 1979.

[44] P. de Mottoni, A. Schiaffino and A. Tesei, On stable space-dependent stationary solutions of a competition system with diffusion, J. Math. Biol., to appear.

[45] C.V. Pao, Bifurcation analysis of a nonlinear diffusion system in reactor dynamics, Applicable Analysis, 9 (1979) 107-119.

[46] C.V. Pao, On nonlinear reaction-diffusion systems, J. Math. Anal. Appl. 87 (1982) 165-198.

[47] L.A. Peletier, A nonlinear eigenvalue problem occuring in population genetics, in Proceedings of Besancon Conference on Nonlinear Analysis, June 1977.

[48] I. Prigogine and P. Glansdorff, Structure, Stabilité et Fluctuations, Masson, Paris 1971.

[49] M.H. Protter and H. Weinberger, Maximum Principles in Differential Equations, Prentice-Hall, Englewood Cliffs, New Jersey 1967.

[50] M. Rascle, Sur une équation integro-differentielle nonlineare issue de la biologie, J. Diff. Eq., 32 (1979) 420-453.

[51] J. Rauch and J. Smoller, Qualitative theory of the FitzHugh-Nagumo equations, Adv. in Math., 27 (1978) 12-44.

[52] M. Reed and B. Simon, Fourier Analysis, Self-Adjointness, Academic Press, New York, San Francisco, London 1975.

[53] F. Rothe, Some analytical results about a simple reaction-diffusion system for morphogenesis, J. Math. Biol., 7 (1979) 375-384.

[54] F.Rothe, Asymptotic behavior of the solutions of the Fisher equation, in Biological Growth and Spread, Mathematical Theories and Applications, Proceedings, Heidelberg 1979, Lecture Notes in Biomathematics 38, Springer-Verlag, New York 1980.

[55] F. Rothe, Solutions for systems of nonlinear elliptic equations with nonlinear boundary conditions, Math. Meth. Appl. Sci., 1 (1979) 545-553.

[56] F. Rothe, Uniform bounds from bounded L_p-functionals in reaction--diffusion equations, J. Diff. Eq., 45 (1982) 207-233.

[57] F. Rothe, Asymptotic behavior of a nonhomogeneous predator-prey system with one diffusing and one sedentary species, Math. Meth. Appl. Sci., 5 (1983) 40-67.

[58] E.T.Rumble and W.E. Kastenberg, On the application of eigenfunction expansion to problems in nonlinear space-time reactor dynamics, Nucl. Sci. Eng., 49 (1972) 172-187.

[59] P.E. Sobolevski, Equations of parabolic type in a Banach space, Am. Math. Soc. Transl., Series 2, 49 (1966) 1-62.

[60] H.P. Schwan, Biological Engineering, Inter-University Electronics Series 9, McGraw-Hill Book Company, New York 1969.

[61] I.E. Segal and R.A. Kunze, Integrals and Operators, Grundlehren der math. Wissenschaften 228, Springer-Verlag, Berlin 1978.

[62] H.B. Stewart, Generation of analytic semigroups by strongly elliptic operators under general boundary conditions, Trans. Amer. Math. Soc., 259 (1980) 299-310.

[63] I. Takagi, Stability of bifurcating solutions of the Gierer-Meinhardt system, Tohoku Math. J., 31 (1979) 221-246.

[64] A.M. Turing The chemical basis of morphogenesis, Phil. Trans. Royal Soc., B237 (1952) 37-72.

[65] W. von Wahl, Klassische Lösbarkeit im Großen für nichtlineare parabolische Systeme und das Verhalten der Lösungen für $t \to \infty$, Nachrichten der Akademie der Wissenschaften in Göttingen II. Mathematisch-Physikalische Klasse, (1981) 131-177.

[66] H.F. Weinberger, Invariant sets for weakly coupled parabolic and elliptic systems, Rendiconti di Mat., 8, Series VI (1975) 295-310.

[67] F.B. Weissler, Local existence and nonexistence for semilinear parabolic equations in L_p, Indiana Univ. J., 29 (1980) 79-102.

[68] F.B. Weissler, Existence and nonexistence of global solutions for a semilinear heat equation, Israel J. Math., 38 (1981) 29-40.

[69] S.A. Williams and P.L. Chow, Nonlinear reaction-diffusion models for interacting populations, J. Math. Anal. Appl., 62 (1978) 157-169.

Index

Vol. 926: Geometric Techniques in Gauge Theories. Proceedings, 1981. Edited by R. Martini and E.M.de Jager. IX, 219 pages. 1982.

Vol. 927: Y. Z. Flicker, The Trace Formula and Base Change for GL (3). XII, 204 pages. 1982.

Vol. 928: Probability Measures on Groups. Proceedings 1981. Edited by H. Heyer. X, 477 pages. 1982.

Vol. 929: Ecole d'Eté de Probabilités de Saint-Flour X – 1980. Proceedings, 1980. Edited by P.L. Hennequin. X, 313 pages. 1982.

Vol. 930: P. Berthelot, L. Breen, et W. Messing, Théorie de Dieudonné Cristalline II. XI, 261 pages. 1982.

Vol. 931: D.M. Arnold, Finite Rank Torsion Free Abelian Groups and Rings. VII, 191 pages. 1982.

Vol. 932: Analytic Theory of Continued Fractions. Proceedings, 1981. Edited by W.B. Jones, W.J. Thron, and H. Waadeland. VI, 240 pages. 1982.

Vol. 933: Lie Algebras and Related Topics. Proceedings, 1981. Edited by D. Winter. VI, 236 pages. 1982.

Vol. 934: M. Sakai, Quadrature Domains. IV, 133 pages. 1982.

Vol. 935: R. Sot, Simple Morphisms in Algebraic Geometry. IV, 146 pages. 1982.

Vol. 936: S.M. Khaleelulla, Counterexamples in Topological Vector Spaces. XXI, 179 pages. 1982.

Vol. 937: E. Combet, Intégrales Exponentielles. VIII, 114 pages. 1982.

Vol. 938: Number Theory. Proceedings, 1981. Edited by K. Alladi. IX, 177 pages. 1982.

Vol. 939: Martingale Theory in Harmonic Analysis and Banach Spaces. Proceedings, 1981. Edited by J.-A. Chao and W.A. Woyczyński. VIII, 225 pages. 1982.

Vol. 940: S. Shelah, Proper Forcing. XXIX, 496 pages. 1982.

Vol. 941: A. Legrand, Homotopie des Espaces de Sections. VII, 132 pages. 1982.

Vol. 942: Theory and Applications of Singular Perturbations. Proceedings, 1981. Edited by W. Eckhaus and E.M. de Jager. V, 363 pages. 1982.

Vol. 943: V. Ancona, G. Tomassini, Modifications Analytiques. IV, 120 pages. 1982.

Vol. 944: Representations of Algebras. Workshop Proceedings, 1980. Edited by M. Auslander and E. Lluis. V, 258 pages. 1982.

Vol. 945: Measure Theory. Oberwolfach 1981, Proceedings. Edited by D. Kölzow and D. Maharam-Stone. XV, 431 pages. 1982.

Vol. 946: N. Spaltenstein, Classes Unipotentes et Sous-groupes de Borel. IX, 259 pages. 1982.

Vol. 947: Algebraic Threefolds. Proceedings, 1981. Edited by A. Conte. VII, 315 pages. 1982.

Vol. 948: Functional Analysis. Proceedings, 1981. Edited by D. Butković, H. Kraljević, and S. Kurepa. X, 239 pages. 1982.

Vol. 949: Harmonic Maps. Proceedings, 1980. Edited by R.J. Knill, M. Kalka and H.C.J. Sealey. V, 158 pages. 1982.

Vol. 950: Complex Analysis. Proceedings, 1980. Edited by J. Eells. IV, 428 pages. 1982.

Vol. 951: Advances in Non-Commutative Ring Theory. Proceedings, 1981. Edited by P.J. Fleury. V, 142 pages. 1982.

Vol. 952: Combinatorial Mathematics IX. Proceedings, 1981. Edited by E. Billington, S. Oates-Williams, and A.P. Street. XI, 443 pages. 1982.

Vol. 953: Iterative Solution of Nonlinear Systems of Equations. Proceedings, 1982. Edited by R. Ansorge, Th. Meis, and W. Törnig. VII, 202 pages. 1982.

Vol. 954: S.G. Pandit, S.G. Deo, Differential Systems Involving Impulses. VII, 102 pages. 1982.

Vol. 955: G. Gierz, Bundles of Topological Vector Spaces and Their Duality. IV, 296 pages. 1982.

Vol. 956: Group Actions and Vector Fields. Proceedings, 1981. Edited by J.B. Carrell. V, 144 pages. 1982.

Vol. 957: Differential Equations. Proceedings, 1981. Edited by D.G. de Figueiredo. VIII, 301 pages. 1982.

Vol. 958: F.R. Beyl, J. Tappe, Group Extensions, Representations, and the Schur Multiplicator. IV, 278 pages. 1982.

Vol. 959: Géométrie Algébrique Réelle et Formes Quadratiques, Proceedings, 1981. Edité par J.-L. Colliot-Thélène, M. Coste, L. Mahé, et M.-F. Roy. X, 458 pages. 1982.

Vol. 960: Multigrid Methods. Proceedings, 1981. Edited by W. Hackbusch and U. Trottenberg. VII, 652 pages. 1982.

Vol. 961: Algebraic Geometry. Proceedings, 1981. Edited by J.M. Aroca, R. Buchweitz, M. Giusti, and M. Merle. X, 500 pages. 1982.

Vol. 962: Category Theory. Proceedings, 1981. Edited by K.H. Kamps, D. Pumplün, and W. Tholen, XV, 322 pages. 1982.

Vol. 963: R. Nottrot, Optimal Processes on Manifolds. VI, 124 pages. 1982.

Vol. 964: Ordinary and Partial Differential Equations. Proceedings, 1982. Edited by W.N. Everitt and B.D. Sleeman. XVIII, 726 pages. 1982.

Vol. 965: Topics in Numerical Analysis. Proceedings, 1981. Edited by P.R. Turner. IX, 202 pages. 1982.

Vol. 966: Algebraic K-Theory. Proceedings, 1980, Part I. Edited by R.K. Dennis. VIII, 407 pages. 1982.

Vol. 967: Algebraic K-Theory. Proceedings, 1980. Part II. VIII, 409 pages. 1982.

Vol. 968: Numerical Integration of Differential Equations and Large Linear Systems. Proceedings, 1980. Edited by J. Hinze. VI, 412 pages. 1982.

Vol. 969: Combinatorial Theory. Proceedings, 1982. Edited by D. Jungnickel and K. Vedder. V, 326 pages. 1982.

Vol. 970: Twistor Geometry and Non-Linear Systems. Proceedings, 1980. Edited by H.-D. Doebner and T.D. Palev. V, 216 pages. 1982.

Vol. 971: Kleinian Groups and Related Topics. Proceedings, 1981. Edited by D.M. Gallo and R.M. Porter. V, 117 pages. 1983.

Vol. 972: Nonlinear Filtering and Stochastic Control. Proceedings, 1981. Edited by S.K. Mitter and A. Moro. VIII, 297 pages. 1983.

Vol. 973: Matrix Pencils. Proceedings, 1982. Edited by B. Kågström and A. Ruhe. XI, 293 pages. 1983.

Vol. 974: A. Draux, Polynômes Orthogonaux Formels – Applications. VI, 625 pages. 1983.

Vol. 975: Radical Banach Algebras and Automatic Continuity. Proceedings, 1981. Edited by J.M. Bachar, W.G. Bade, P.C. Curtis Jr., H.G. Dales and M.P. Thomas. VIII, 470 pages. 1983.

Vol. 976: X. Fernique, P.W. Millar, D.W. Stroock, M. Weber, Ecole d'Eté de Probabilités de Saint-Flour XI – 1981. Edited by P.L. Hennequin. XI, 465 pages. 1983.

Vol. 977: T. Parthasarathy, On Global Univalence Theorems. VIII, 106 pages. 1983.

Vol. 978: J. Ławrynowicz, J. Krzyż, Quasiconformal Mappings in the Plane. VI, 177 pages. 1983.

Vol. 979: Mathematical Theories of Optimization. Proceedings, 1981. Edited by J.P. Cecconi and T. Zolezzi. V, 268 pages. 1983.

Vol. 980: L. Breen. Fonctions thêta et théorème du cube. XIII, 115 pages. 1983.

Vol. 981: Value Distribution Theory. Proceedings, 1981. Edited by I. Laine and S. Rickman. VIII, 245 pages. 1983.

Vol. 982: Stability Problems for Stochastic Models. Proceedings, 1982. Edited by V. V. Kalashnikov and V. M. Zolotarev. XVII, 295 pages. 1983.

Vol. 983: Nonstandard Analysis-Recent Developments. Edited by A. E. Hurd. V, 213 pages. 1983.

Vol. 984: A. Bove, J. E. Lewis, C. Parenti, Propagation of Singularities for Fuchsian Operators. IV, 161 pages. 1983.

Vol. 985: Asymptotic Analysis II. Edited by F. Verhulst. VI, 497 pages. 1983.

Vol. 986: Séminaire de Probabilités XVII 1981/82. Proceedings. Edited by J. Azéma and M. Yor. V, 512 pages. 1983.

Vol. 987: C. J. Bushnell, A. Fröhlich, Gauss Sums and p-adic Division Algebras. XI, 187 pages. 1983.

Vol. 988: J. Schwermer, Kohomologie arithmetisch definierter Gruppen und Eisensteinreihen. III, 170 pages. 1983.

Vol. 989: A. B. Mingarelli, Volterra-Stieltjes Integral Equations and Generalized Ordinary Differential Expressions. XIV, 318 pages. 1983.

Vol. 990: Probability in Banach Spaces IV. Proceedings, 1982. Edited by A. Beck and K. Jacobs. V, 234 pages. 1983.

Vol. 991: Banach Space Theory and its Applications. Proceedings, 1981. Edited by A. Pietsch, N. Popa and I. Singer. X, 302 pages. 1983.

Vol. 992: Harmonic Analysis, Proceedings, 1982. Edited by G. Mauceri, F. Ricci and G. Weiss. X, 449 pages. 1983.

Vol. 993: R. D. Bourgin, Geometric Aspects of Convex Sets with the Radon-Nikodým Property. XII, 474 pages. 1983.

Vol. 994: J.-L. Journé, Calderón-Zygmund Operators, Pseudo-Differential Operators and the Cauchy Integral of Calderón. VI, 129 pages. 1983.

Vol. 995: Banach Spaces, Harmonic Analysis, and Probability Theory. Proceedings, 1980–1981. Edited by R.C. Blei and S.J. Sidney. V, 173 pages. 1983.

Vol. 996: Invariant Theory. Proceedings, 1982. Edited by F. Gherardelli. V, 159 pages. 1983.

Vol. 997: Algebraic Geometry – Open Problems. Edited by C. Ciliberto, F. Ghione and F. Orecchia. VIII, 411 pages. 1983.

Vol. 998: Recent Developments in the Algebraic, Analytical, and Topological Theory of Semigroups. Proceedings, 1981. Edited by K.H. Hofmann, H. Jürgensen and H. J. Weinert. VI, 486 pages. 1983.

Vol. 999: C. Preston, Iterates of Maps on an Interval. VII, 205 pages. 1983.

Vol. 1000: H. Hopf, Differential Geometry in the Large, VII, 184 pages. 1983.

Vol. 1001: D.A. Hejhal, The Selberg Trace Formula for PSL(2, ℝ). Volume 2. VIII, 806 pages. 1983.

Vol. 1002: A. Edrei, E.B. Saff, R.S. Varga, Zeros of Sections of Power Series. VIII, 115 pages. 1983.

Vol. 1003: J. Schmets, Spaces of Vector-Valued Continuous Functions. VI, 117 pages. 1983.

Vol. 1004: Universal Algebra and Lattice Theory. Proceedings, 1982. Edited by R. S. Freese and O. C. Garcia. VI, 308 pages. 1983.

Vol. 1005: Numerical Methods. Proceedings, 1982. Edited by V. Pereyra and A. Reinoza. V, 296 pages. 1983.

Vol. 1006: Abelian Group Theory. Proceedings, 1982/83. Edited by R. Göbel, L. Lady and A. Mader. XVI, 771 pages. 1983.

Vol. 1007: Geometric Dynamics. Proceedings, 1981. Edited by J. Palis Jr. IX, 827 pages. 1983.

Vol. 1008: Algebraic Geometry. Proceedings, 1981. Edited by J. Dolgachev. V, 138 pages. 1983.

Vol. 1009: T. A. Chapman, Controlled Simple Homotopy Theory and Applications. III, 94 pages. 1983.

Vol. 1010: J.-E. Dies, Chaînes de Markov sur les permutations. IX, 226 pages. 1983.

Vol. 1011: J.M. Sigal. Scattering Theory for Many-Body Quantum Mechanical Systems. IV, 132 pages. 1983.

Vol. 1012: S. Kantorovitz, Spectral Theory of Banach Space Operators. V, 179 pages. 1983.

Vol. 1013: Complex Analysis – Fifth Romanian-Finnish Seminar. Part 1. Proceedings, 1981. Edited by C. Andreian Cazacu, N. Boboc, M. Jurchescu and I. Suciu. XX, 393 pages. 1983.

Vol. 1014: Complex Analysis – Fifth Romanian-Finnish Seminar. Part 2. Proceedings, 1981. Edited by C. Andreian Cazacu, N. Boboc, M. Jurchescu and I. Suciu. XX, 334 pages. 1983.

Vol. 1015: Equations différentielles et systèmes de Pfaff dans le champ complexe – II. Seminar. Edited by R. Gérard et J.P. Ramis. V, 411 pages. 1983.

Vol. 1016: Algebraic Geometry. Proceedings, 1982. Edited by M. Raynaud and T. Shioda. VIII, 528 pages. 1983.

Vol. 1017: Equadiff 82. Proceedings, 1982. Edited by H. W. Knobloc and K. Schmitt. XXIII, 666 pages. 1983.

Vol. 1018: Graph Theory, Łagów 1981. Proceedings, 1981. Edite M. Borowiecki, J. W. Kennedy and M.M. Sysło. X, 289 pages. 1

Vol. 1019: Cabal Seminar 79–81. Proceedings, 1979–81. Edite A. S. Kechris, D. A. Martin and Y. N. Moschovakis. V, 284 pages. 1

Vol. 1020: Non Commutative Harmonic Analysis and Lie Gro Proceedings, 1982. Edited by J. Carmona and M. Vergne. V, pages. 1983.

Vol. 1021: Probability Theory and Mathematical Statistics. ceedings, 1982. Edited by K. Itô and J.V. Prokhorov. VIII, 747 pa 1983.

Vol. 1022: G. Gentili, S. Salamon and J.-P. Vigué. Geometry Sem "Luigi Bianchi", 1982. Edited by E. Vesentini. VI, 177 pages. 198

Vol. 1023: S. McAdam, Asymptotic Prime Divisors. IX, 118 pa e 1983.

Vol. 1024: Lie Group Representations I. Proceedings, 1982–198 Edited by R. Herb, R. Lipsman and J. Rosenberg. IX, 369 pages. 198

Vol. 1025: D. Tanré, Homotopie Rationnelle: Modèles de Ch Quillen, Sullivan. X, 211 pages. 1983.

Vol. 1026: W. Plesken, Group Rings of Finite Groups Over p-a Integers. V, 151 pages. 1983.

Vol. 1027: M. Hasumi, Hardy Classes on Infinitely Connected mann Surfaces. XII, 280 pages. 1983.

Vol. 1028: Séminaire d'Analyse P. Lelong – P. Dolbeault – H. Sk Années 1981/1983. Edité par P. Lelong, P. Dolbeault et H. Skod VIII, 328 pages. 1983.

Vol. 1029: Séminaire d'Algèbre Paul Dubreil et Marie-Paule Malliav Proceedings, 1982. Edité par M.-P. Malliavin. V, 339 pages. 1983

Vol. 1030: U. Christian, Selberg's Zeta-, L-, and Eisensteinserie XII, 196 pages. 1983.

Vol. 1031: Dynamics and Processes. Proceedings, 1981. Edite Ph. Blanchard and L. Streit. IX, 213 pages. 1983.

Vol. 1032: Ordinary Differential Equations and Operators. Pr ceedings, 1982. Edited by W. N. Everitt and R. T. Lewis. XV, 521 page 1983.

Vol. 1033: Measure Theory and its Applications. Proceedings, 198 Edited by J. M. Belley, J. Dubois and P. Morales. XV, 317 pages. 198

Vol. 1034: J. Musielak, Orlicz Spaces and Modular Spaces. 222 pages. 1983.